Protein Chromatography

Protein Chromatography

Mohamed A. Selmy, MD PhD

ARCLER PRESS

www.arclerpress.com

Protein Chromatography

Mohamed A. Selmy, MD PhD

Arcler Press
2010 Winston Park Drive,
2nd Floor
Oakville, ON L6H 5R7
Canada
www.arclerpress.com
Tel: 001-289-291-7705
001-905-616-2116
Fax: 001-289-291-7601
Email: orders@arclereducation.com

ISBN: 978-1-77361-158-7 (Hardcover)

ABOUT THE AUTHOR

Mohamed A. Selmy, MD PhD

Mohamed Selmy is a physician scientist and lecturer of Medical Biochemistry and Molecular Biology. He received his medical degree in 2004 from the faculty of medicine Suez Canal University in Egypt. He received his scientific training through many fellowships in United States, Spain and Italy. He received his PhD in Medical Biochemistry and Cancer biology in 2017 from the faculty of medicine Suez Canal University in Egypt. His fields of expertise comprises; cancer biology, cell signaling and molecular medicine. His work is continuously published in international peer reviewed journals.

TABLE OF CONTENTS

LIST OF ABBREVIATIONS

%	Percentage
%CV	Coefficient Variation
1-D	One-dimensional 2-D two-dimensional
2-DE	Two-dimensional gel Electrophoresis
AC	Affinity Chromatography
ANP	Squeous Normal Phase
API	Atmospheric Pressure Lonization
AQ	Aqueous
BSA	Bovine Serum Albumin
CAN	Acetonitrile
CC	Calibration Curve
CF	Chromato Focussing
CV	Coefficient of Variation
Da	Dalton
EF	Isoelectric Focusing
FPLC	Fast Protein Liquid Chromatography
GC	Fas Chromatography
GC-MS	Gas Chromatography-Mass Spectrometry
HPTLC	High-Performance Thin Layer Chromtography
HPLC	High Performance Liquid Chromatography
IEC	Ion Exchange Chromatography I
LC	Liquid Chromatography

LC-IR	Liquid Chromatography with Infrared
LC-MS	Liquid Chromatography with Mass Spectrometry
LLE	Liquid – Liquid Extraction
LLOQ	Lower Limit of Quantification
LLOQQC	Lower Limit of Quantification Quality Control
LOD	Limit of Detection
LOQ	Limit of Quantification
LQC	Lower Quality Control
MAC	Membrane Affinity Chromatography
MALDI-MS	Matrix-Assisted Laser Desorption/Ionization-Mass Spectrometry
MEKC	Micellar Electrokinetic Capillary Chromatography
mg :	Milligram
μg	Microgram
μL	Micro liter
min	Minute (S)
mL	Milliliter
mM	Millimolar
MS	Mass Spectrometry
MS/MS	Tandem Mass Spectrometry
PDA	Photodiode Array
PVP	polyvinylpyrrolidone
r	Correlation Coefficient
RP	Reversed Phase
RP-HPLC	Reverse Phase Liquid Chromatography
RP-UPLC	Reverse Phase Ultra Performance Liquid Chromatography
RRT	Related retention time RS Related substances

RSD	Relative Standard Deviation
RT	Retention Time
S/N	Signal to Noise
SD	Standard Deviation
SEC	Size Exclusion Chromatography
SPE	Solid Phase Extraction
STD	Calibration curve standard
TFA	Trifluoroacetic Acid
TIMS	Thermal Ionization Mass Spectrometry
TLC	Thin-layer chromatography
TOF	Time of Flight
UHPLC	Ultra high performance liquid chromatography
ULOQ	Upper limit of quantification
UPLC	Ultra performance liquid chromatography
USFDA	United states-food and drug administration
USP	United states pharmacopeia
UV	Ultraviolet
UV-VIS	Ultraviolet-visible
v/v	Volume by volume
WHO	World health organization

LIST OF FIGURES

LIST OF TABLES

PREFACE

The separation and isolation of components of a mixture have always captured the fancy of man. Right from the isolation of salt from sea water to the separation of sand and sugar or advanced separation of a complex mixture, several techniques have been invented. Mikhail Semenovich Tswett set the ball rolling with the principle of separation of plant pigments using filter paper and solvent. Today, the technique stands with several branches and applications. There are several types of chromatography on the basis of their principles while the following are the major varieties.

Column chromatography, Ion-exchange chromatography, Gel-permeation (molecular sieve) chromatography, Affinity chromatography, Paper chromatography, Thin-layer chromatography, Gas chromatography, Dye-ligand chromatography, Hydrophobic interaction chromatography, Pseudoaffinity chromatography, High-pressure liquid chromatography, Super critical fluid chromatography and Flash chromatography.

The techniques have been applied to the separation of compounds, characterization of new molecules, analysis of complex mixtures with both qualitative and quantitative applications. Chromatography offers assistance to all: from high school labs to college experiments to research institutes and the ever-developing biotechnology industry. This book offers collection of the principles, techniques and applications of Protein chromatography with case studies to make the reading more purposeful.

Author

CHAPTER 1

BASICS OF CHROMATOGRAPHY

INTRODUCTION

According to IUPAC, the technique of chromatography is defined as, "a physical method of separation in which the components to be separated are distributed between two phases, one of which is stationary (stationary phase) while the other (the mobile phase) moves in a definite direction" (1).

The word chromatography has two words with roots in Greek language: chroma means color and graphein means to write. As separated components of a mixture appeared as rings of several colors, the term can be translated as "color writing". The term can also be attributed a meaning based on the surname of its inventor: Tswett, that means color (2).

A GLIMPSE INTO THE HISTORY OF CHROMATOGRAPHY

The origins of chromatography can be traced back to as early as 1903 when Mikhail Semenovich Tswett delivered a talk at a meeting of the

Biological Section of the Warsaw Society of Natural Sciences, that revolved around, "On a New Category of Adsorption Phenomena and Their Application to Biochemical Analysis.". The lecture involved elucidation of works involving the separation of pigments of leaves that was later developed as chromatography.

Tswett worked on separation of plant pigments from leaves using solvents: he reported that ethanol or acetone could elicit the complete set of pigments while ligroin or petroleum ether could achieve partial extraction. With a view to mimic the structure of leaves, he narrowed down on filter paper that has the same framework: cellulose. Following extraction of leaf pigments from ethanol, and drying the dipping of the paper in ligroin and ethanol caused resolution of all the pigments.

He then proceeded to check the effects of various adsorbents for the extraction of pigments: with inulin, calcium carbonate, and alumina ranking the best. These adsorbents caused the separation of the chlorophyll into green and yellow bands with sometimes the formation of new bands.

The separation of chlorophylls, carotenes and xanthophylls was reported by following the technique of adsorption and extraction. The color of the band and the absorption spectrum on UV was used for the analysis of the compounds.

Subsequent refinement and development in the technique using several solvents and columns, he wrote the following words in his paper in 1906:

"Like light rays in the spectrum, the different components of a pigment mixture, obeying a law, are resolved on the calcium carbonate column and then can be qualitatively and quantitatively determined. I call such a preparation a chromatogram and the corresponding method the chromatographic method" (3,4).

In another work by Leroy S. Palmer, the technique of Tswett's work was used for the separation of pigments in food products. He then worte a book entitled" Carotinoids and Related Pigments: The Chromolipids " in 1922 (2).

Chromatography then resurfaced in 1931 when Edgar Lederer referred to the above-mentioned book and applied *the* technique with success for separation of carotenoids in egg yolk (5).

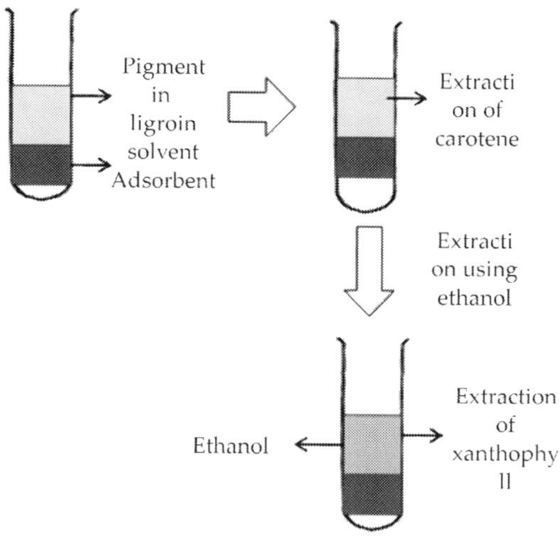

Figure 1.1: Scheme of Tswett's initial experiment

Following such developments, the technique of process chromatography appeared. Tswett's work gave rise to research for several matrices and packing columns ranging from styrene-divinyl benzene hydrophobic resins to cellulose ion exchangers and cross-linked dextrans:Sephadex), followed by agarose and polyacrylamide. The development of affinity chromatography and the use of IgG purification using Protein A adsorbents brought in a new era in the field of research. This was followed by the development of reversed-phase partition chromatography with the use of cellulose acetate treated with solvent as a continuous stationary phase.

The technique of hydrophobic interaction chromatography used a polar mobile phase with polar ligands attached to a neutral base. Such techniques based on reverse phase led to the dawn of high-performance liquid chromatography (HPLC) that is widely employed today. The very first attempt of scale up of such technologies is credited to Pharmacia Fine Chemicals for the creation of "Gel Filters" or Sephadex G-25 in stainless steel columns.

A novel application of chromatography in the field of biotechnology was the single peak insulin by Eli Lilly that used gel filtration on Sephadex G-50 to produce the hormone that appeared as a single peak, reviewed by John Curling (6).

PRINCIPLE OF CHROMATOGRAPHY

The technique of chromatography relies on *resolution or separation of sample* compounds or analytes for subsequent analysis or use. Based on the principles of adsorption or partition, analytes differ in the affinity towards either the stationary phase or mobile phase following which they are detected.

In the mechanism of *adsorption*, the polarity or non-polarity play key roles in the separation of compounds. With a polar stationary phase and polar analyte, the sample traverses the column in a slow manner as compared to non-polar components to be eluted out last.

In *partition*, the solubility of the fractions of mixtures in different media aids in the calculation of a partition coefficient. Based on the preferential partitioning of a sample into a phase, separation is achieved.

Samples in a mixture are allowed to pass through a surface containing *stationary* or fixed molecules as a sample moves with the help of a mobile or phase. On the basis of the principles described above: *adsorption or partition, the samples are separated or resolved* (7). The differences in the properties of the samples causes certain analytes to remain in the stationary phase for a longer duration to come out later while certain analytes enter the mobile phase and exit from the system more rapidly (8).

The major elements of a chromatographic separation are:

Stationary phase: This is a solid or liquid that is coated or applied or adsorbed on the surface of a support material.

Mobile phase: It is a liquid or gas that flows through the system.

Sample: The analytes or mixture that is under study for identification or separation.

A solid or liquid phase applied on a solid surface constitutes the stationary phase in chromatography, while a liquid or gas that flows over the stationary phase is a mobile phase. When a liquid is used as mobile phase: it is termed as liquid chromatography (LC) while gas chromatography (GC) is constituted by a gas as a mobile phase. The former finds application for thermally labile and non-volatile samples while the latter finds use in separation of volatile liquids, gases (9, 10).

GLOSSARY OF TERMS USED IN CHROMATOGRAPHY

Chromatography

Chromatography is a physical method of separation in which the components to be separated are distributed between two phases, one of which is stationary while the other pursues a path in a single direction and is called mobile phase.

Chromatogram

A presentation, mostly graphical, to measure the concentration of an analyte in the effluent as a measure of time. It can also be applied to the surface on which an experiment is conducted for planar chromatography.

DETAILS OF CHROMATOGRAMS

Differential Chromatogram

It is a chromatogram recorded with a differential detector

Integral Chromatogram

It is a chromatogram recorded with an integral detector

Starting Point or Line

It is a point or line on a chromatographic system where an analyte is applied.

Spot

It is a zone especially of circular nature in paper and thin-layer chromatography.

Spot Diameter

It is the width of a sample component spot.

Baseline

It is the portion of the chromatogram where the readings of the mobile phase flowing in a column are recorded.

Peak

It is the portion of a differential chromatogram where the exit or elution of an analyte is recorded by the detector.

Peak Base

It is an interpolation of the baseline between the peak boundaries.

Peak Area

It refers to the area that is encompassed peak base and the peak.

Peak Maximum

The point with maximum value of the distance from the base and is quantified along the same axis used for representation of detector response.

Peak Height

It is the distance between the peak maximum and the peak base, measured in a direction parallel to the axis that represents detector response.

Peak-Widths

They are representative of retention dimensions (time or volume) parallel to the baseline.

Peak-Width at Base

It constitutes a section of the peak base that is subject to interception by tangents on the peak sides.

Chromatograph (verb)

Refers to achieve separation using chromatography.

Chromatograph (noun)

Refers to the instruments and apparatus to perform a chromatographic separation.

Stationary Phase

It comprises a phase of the two that constitute a chromatographic system, that can be a solid, a gel or a liquid. In the case of the former, there is a chance of contribution to the separation while in the case of the latter it can be bonded to it: by immobilization *(Immobilized Phase)* or chemically bonded *(Bonded Phase).*

Bonded Phase

It is a type of a stationary phase that involves covalent bonding of covalent nature of the phase and tsupport particles or to the inner side of the column.

Immobilized Phase

It is a type of a stationary phase which is involves immobilization on the support particles, or on the inner end of the columns.

Mobile Phase

The second phase that constitutes a chromatographic system that percolates or moves across the stationary phase in a definite direction. The phase may be a gas *(Gas Chromatography)* or a liquid *(Liquid Chromatography)* or a supercritical fluid *(Supercritical-Fluid Chromatography).* In gas chromatography, the term mobile phase may be replaced with the expression *Carrier Gas.* In elution chromatography, the term mobile phase may be replaced with the expression *Eluent.*

Elute (verb)

It involves the separation of samples in a chromatography experiment.

TYPES OF ELUTION

Isocratic Analysis

It involves a procedure in which the composition of the mobile phase remains constant throughout the elution process.

Gradient Elution

It involves a procedure in which the composition of the mobile phase is changed continuously or stepwise throughout elution process.

Stepwise Elution

It involves a procedure in which the composition of the mobile phase is changed through a series of steps throughout elution process.

This term has superseded term *Develop* used earlier in planar chromatography.

Effluent

It refers to the components that exit the column.

Sample Components

The chemically pure constituents of the sample. They may be unretained *(Le.,* not delayed) by the stationary phase, partially retained *(Le.,* eluted at different times) or retained permanently.

Solute

A term that concerns the sample components or analytes in partition chromatography.

Solvent

A term that refers to the liquid stationary phase used in partition chromatography. The usage of "solvent: in place of mobile phase is not recommended in liquid chromatography.

Zone

A region in the chromatographic area that is the location where an analyte is present. It can also be called as a band.

Flow Rate

It is the volume of mobile phase traversing through the chromatographic column in unit time.

1.7 Apparatus in Chromatography

Pump

It finds application in liquid chromatography and is an apparatus that delivers the mobile phase at a controlled flow-rate to the separation system.

The pumps are of the following types:

Syringe Pumps

They possess a piston that moves at a controlled rate within a smooth cylinder to effect displacement of the mobile phase.

Reciprocating Pumps

They possess a single or multiple chamber, to effect displacement of the mobile phase via a reciprocating piston(s) or diaphragm(s).

Pneumatic Pumps

They use gases to effect displacement of the mobile phase using a piston or direct action.

Sample Injector

A device that causes introduction of a liquid, solid or gaseous sample into the mobile phase or the chromatographic system.

They are of several types:

Direct Injector

A device that effects direct introduction of a sample into the mobile-phase stream.

Bypass Injector

A device in which a sample is initially introduced into a chamber and using a system of valves, an instant introduction of the mobile phase stream through the chamber can be caused, to carry the sample to the column. A bypass injector may also be known as a *Valve Injector* or *Sampling Valve.*

On-Column Injector

It effects direct introduction of a sample into the column. In gas chromatography, the on-column injector facilitates application of the liquid sample into the column without initial evaporation.

Flash Vaporizer

A device that causes introduction of a liquid sample into the carrier gas stream with simultaneous evaporation and mixing with the carrier gas before entry into the column, applied in gas chromatography.

Split Injection

It is a type of sample introduction technique used in gas chromatography. The sample is subjected to flash vaporization subsequent to which mixing of the sample with the carrier gas is performed. The stream is split into two fractions, one applied to the column and the other is discarded.

Programmed Temperature Vaporizer (PTV)

It is a type of sample introduction technique used in gas chromatography. The liquid sample is applied, usually with a syringe, into a device similar to a flash vaporizer where the temperature is maintained below the boiling point of the sample components. This is followed by rapid heating of device in a controlled fashion to evaporate the sample into the carrier gas stream. The PTV may also be used in the split mode: in this where, the carrier gas stream containing the evaporated sample components is split into two fractions, one applied to the column and the other is discarded.

Gas *Sampling Valve*

It is a bypass injector that facilitates introduction of a gaseous sample of a given volume into a gas

Chromatographic instrument.

Column Oven

A thermostatically controlled oven containing the column whose temperature can be applied over a wide range.

Fraction Collector

It is a device that recovers fractional volumes of the effluent exiting column.

Detector

It is a device that measures the change in the composition of an analyte that leaves the column by evaluation of its physical or chemical properties. The detectors used in chromatography are as follows:

Based on response:

Differential Detectors

They measure the difference at a given instant in the composition of the eluted samples.

Integral Detectors

These measure the amount of the eluted samples as recorded by a detector.

Based on measurements

Mass-Flow Sensitive Detector

It is a device whose response of which is proportional to the amount of sample component that passes the detector as a function of time.

Concentration-Sensitive Detector

It is a device whose response of which is proportional to the concentration of a sample component.

Based on the samples recorded:

Universal Detector

Barring the mobile phase, every sample is recorded by the system.

Selective Detector

A detector that records a related group of sample components in the eluted samples.

Specific Detector

A detector that records a single sample analyte or to a limited number of analytes that possess similar chemical characteristics.

Internal Standard

It is a compound added to a sample in known concentration to permit the qualitative identification and/or quantitative determination of analytes.

External Standard

It is a compound that allows qualitative identification and/or quantitative determination of the sample components. It is a standard sample whose concentration and volume are known to be analysed separately from the unknown sample under identical conditions.

Marker

It is a reference substance assayed along with the sample to aid in the identification of the analytes

METHODS IN CHROMATOGRAPHY

Chromatography classification based on application:

Frontal Chromatography

It is a procedure in which a liquid or gas sample is applied in a continuous fashion into the chromatographic system.

Displacement Chromatography

It is a procedure in which the mobile phase contains a substance called *Displacer* that is more strongly retained than the analytes.

Elution Chromatography

It is a procedure in which the mobile phase is in a continuous fashion through or along thechromatographic system.

CHROMATOGRAPHY CLASSIFICATION BASED ON SHAPE COLUMN CHROMATOGRAPHY

It is a separation technique in which the stationary phase is within a tube/ column. The particles of the solid stationary phase or the support coated with a liquid stationary phase can occupy the entire volume of the tube *(Packed Column)* or is applied on inside tube wall leaving a central path to allow the flow of mobile phase *(Open-TubularColumn)*.

Planar Chromatography/ Open-Bed Chromatography

It is a separation technique in which the stationary phase is applied on a plane. The plane can

Take the form of a paper that doubles up as substance as the stationary phase (*Paper Chromatography,* PC) or a layer of solid particles applied on say., a glass plate *(Thin Layer Chromatography,* TLC).

CHROMATOGRAPHY CLASSIFICATION BASED ON THE PHASE OF THE COMPONENTS

Chromatographic techniques can be classified using physical state of *both* phases used. There are the following types:

- Gas-liquid chromatography (GLC)
- Gas-solid chromatography (GSC)
- Liquid-liquid chromatography (LLC)
- Liquid-solid chromatography (LSC)

Gas Chromatography (GC)

It is a separation technique in which the mobile phase is a gas performed in a column.

Liquid Chromatography (LC)

It is a separation technique in which the mobile phase is a liquid.

Supercritical-Fluid Chromatography (SFC)

It is a separation technique in which the mobile phase is a fluid near or beyond its critical temperature and pressure.

TYPES OF CHROMATOGRAPHY

Adsorption Chromatography

The technique involves variations in adsorption affinities of the sample on surface of an active solid.

Partition Chromatography

The technique involves variations between the solubilities of the sample components in the stationary phase in gas chromatography, or on differences between the solubilities of the components in the mobile and stationary phases, in liquid chromatography.

Ion-Exchange Chromatography

The technique involves differences in the ion exchange affinities of the sample components.

Exclusion chromatography

The technique involves exclusion effects, such as differences in molecular size along with shape or in charge. The term *Size-Exclusion Chromatography* may also be used when separation is based on molecular size. The terms *Gel Filtration* and *Gel-Permeation*

Chromatography (GPC) was earlier applied when gels were used while the term *Ion-Exclusion Chromutography* is specifically used for the separation of aqueous ions.

Afinity Chromatography

The technique involves the unique biological specificity of the analyte and ligand binding for the separation.

OTHER TECHNIQUES

Reversed-Phase Chromatography

It is an elution procedure used in liquid chromatography in which the mobile phase possesses more polarity than the stationary phase.

Normal-Phase Chromatography

It is an elution procedure used in liquid chromatography in which the mobile phase possesses less polarity than the stationary phase.

Two-Dimensional Chromatography

It is a procedure in which parts or all of the separated analytes are subjected to additional stages of separation. It can be achieved by applying eluted samples from a column into another column having based on another technique of separation.

It can also be referred to as *Multi-Dimensional Chromatography.*

In planar chromatography, two-dimensional chromatography involves separation of components in one direction followed by another separation in a direction at right angles to the first one; the two elutions are carried out with different eluting agents.

Isothermal Chromatography

It is a procedure in which the temperature of the column is kept constant during the experiment.

Programmed-Temperature Chromatography (Temperature Programming)

It is a procedure in which the temperature of the column is changed in a controlled fashion during certain parts or the entire separation.

Programmed-Flow Chromatography (Flow Programming)

It is a procedure in which the rate of flow of the mobile phase is changed in a controlled fashion during certain parts or the entire separation.

Programmed-Pressure Chromatography (Pressure Programming)

It is a procedure in which the inlet pressure of the mobile phase is changed changed in a controlled fashion during certain parts or the entire separation.

Reaction Chromatography

It is an approach in which the identities of the sample components are changed between sample introduction and detection purposely. It can be done as follows:

Pyrolysis-Gas Chromatography

It is an offshoot of reaction chromatography in which a sample is subjected to thermal decomposition to simple fragments before utility on the column.

Pre/Post-Column Derivatization

It is another aspect of reaction chromatography wherein the separated pattern additives eluting from the column are subjected to derivatization earlier than their access into the detector. Alternatively, if the process occurs before their entry into the column, it's far referred to as pre-column (preliminary) derivatization.

TERMS USED IN THE CHROMATOGRAPHIC MEDIUM USED

Active Solid

It is a solid that has absorptive/adsorptive properties.

Modified Active Solid

It is an active solid who's the absorptive/adsorptive properties have been modified,

Solid Support

It is a solid that bears the stationary phase and is expected not to contribute to the process of separation.

Binders

These are additives used to keep the solid stationary phase in place on an inactive plate or sheet in thin-layer chromatography.

Gradient Layer

It refers to a chromatographic bed used in thin-layer chromatography in which there is a gradual shift in a particular property.

Impregnation

Refers to modification of the separation properties of the chromatographic bed used in planar chromatography by the use of appropriate additives.

Packing

It is the active solid, stationary liquid along with its solid support, or swollen gel present in a tube. It can either be:

Totally Porous Packing

It involves permeation of stationary phase into each porous particle.

Pellicular Packing

It involves formation of a shell of the stationary phase around an impermeable particle.

Particle Diameter

It is a measure of average diameter of the solid particles.

Pore Radius

It is a measure of average radius of the pores within the solid particles.

Liquid-Phase Loading

It is a terminology used in partition chromatography to express the relative amount of the liquid stationary phase in the column packing.

VARIOUS COLUMNS USED IN CHROMATOGRAPHY

Packed Column

It is a tube composed of a solid packing.

Open-Tubular Column

It is a tube column, possessing a small diameter where a liquid or active solid are coated on the inner tube wall to serve as a stationary phase with an open, unrestricted path for the mobile phase in the centre.

Wall-Coated Open-Tubular (WCOT) Column

In these columns, the liquid stationary phase is applied on the inner wall of the tube.

Porous-Layer Open-Tubular (PLOT) Column

In these columns, there is a porous layer on the inner wall that can be the stationary phase or a support for a liquid stationary phase.

Support-Coated Open-Tubular (SCOT) Column

It is a tube version of a PLOT in which a porous layer consists of support particles.

Capillary Column

It includes columns of a small diameters.

A FEW TECHNICAL DEFINITIONS

Column Volume (V)

It is the geometric volume of the part of the tube that contains the packing and is defined as:

$Vc=AcL$; where Ac is the internal cross-sectional area of the tube and L is the length of the portion packed in the column.

Extra-column Volume

It is the volume between the effective injection point and the effective detection points: mainly injector volume, connections and detector.

Dead-Volume

It is another way to express the extra-column volume. Technically it is the volume in a chromatographic system where the mobile phase does not have access to (IUPAC Nomenclature: L. S. Ettre, 1993).

CLASSIFICATION OF CHROMATOGRAPHY

The classification of techniques in chromatography is of several criteria that include:

When the physical state of the stationary and mobile phases is examined there are 2 categories:

Homogeneous techniques are comprised of a single phase: liquid as stationary phase and mobile phase as liquid. Example: Liquid-liquid chromatography.

Heterogeneous techniques are composed of different states of matter that serve as stationary and mobile phases. Examples: Solid-Gas chromatography, Solid-liquid chromatography, Liquid-Gas chromatography.

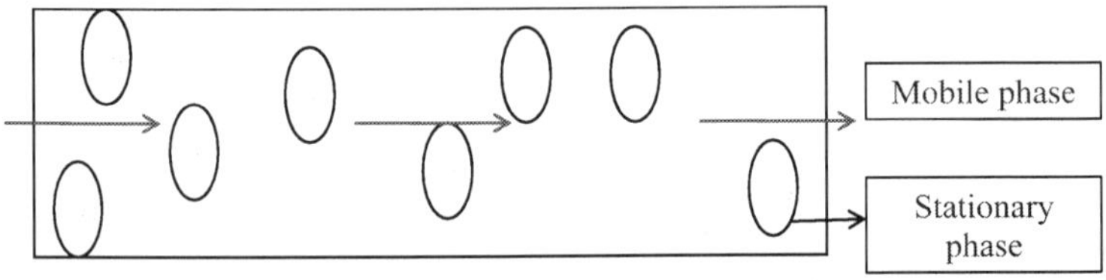

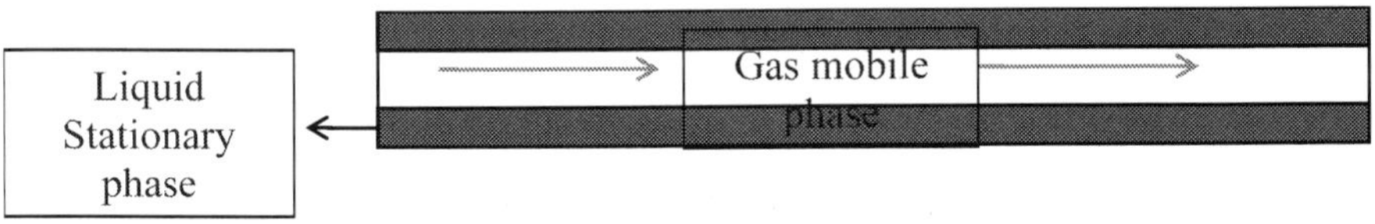

A summary of chromatographic techniques:

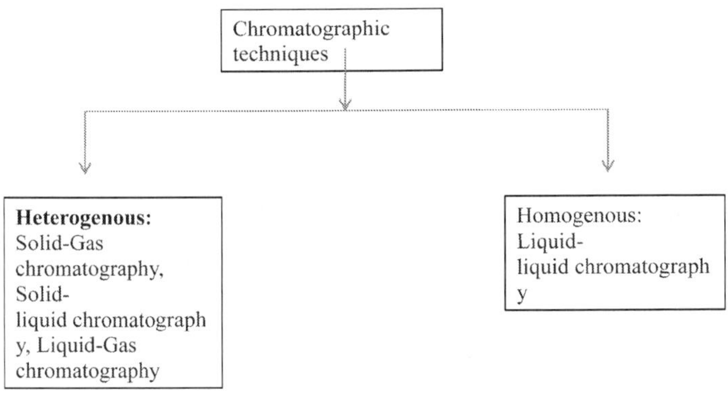

Figure 1.2: Liquid chromatography: The liquid mobile phase is shown in red arrows as it flows through a stationary phase packed in a support

PROCESS OF SEPARATION

When the tenet behind the process of separation is used for classification there are 2 types.

Partition chromatography: This technique uses the variations in the differences in the patterns of partition into the phases. Certain analytes that partition or prefer the stationary phase, remain in the column or move less fast while analytes that partition or prefer the mobile phase move out with it and are released earlier or faster.

Adsorption chromatography: The analytes under test are separated on the basis of their tendency to get adsorbed on the stationary phase rather than move with the mobile phase.

When the tenet behind the experimental purpose is used for classification there are 2 types:

Analytical chromatography: The experiment involves quantification or the identification of the analytes in a sample. This requires a small quantity of sample while following analysis the samples removed from the column are discarded.

Preparative chromatography: The experiment involves the separation and preparation of an analyte for further study or use. It entails a huge quantity of sample while following analysis the samples removed from the column are put to use or study.

When the tenet behind the structure of stationary phase is used for classification there are 2 types:

Column chromatography: The stationary phase is used and packed as a column in which the resolution takes place.

Planar chromatography: A flat dimensioned material is used as the stationary phase. It is shown in the following scheme:

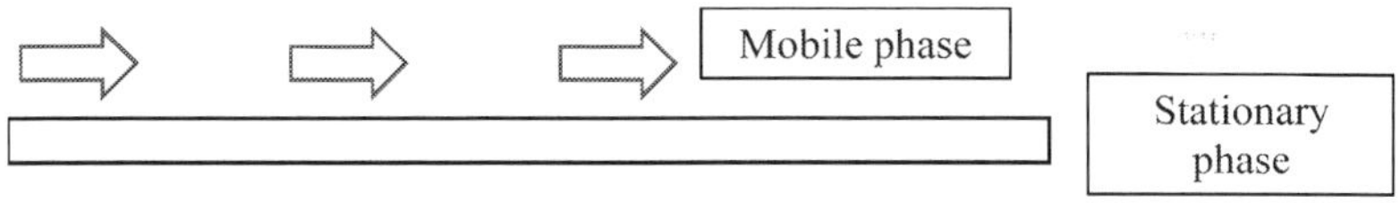

Figure 1.3: The mobile phase is shown in red arrows as it flows over a flat stationary phase

When the tenet behind the polarity is used for classification there are 2 types:

Normal phase chromatography: A polar material is used as stationary phase while a non-polar material finds application as the mobile phase. Analytes that are non-polar partition into the mobile phase and are separated earlier than polar compounds that are retained in the stationary phase.

Reverse phase chromatography: As a reversed technique, a non-polar polar material is used as stationary phase while a polar material finds application as the mobile phase. Materials that have multifold application such as drugs are polar and are tested using this system.

OVERVIEW OF TYPES OF CHROMATOGRAPHY

Due to differences in the various features of sample analytes such as charge, shape, size as well as features of a system such as stationary phase, the following techniques can be applied for chromatography.

Based on the *mode of operation*, the technique of chromatography is of the following types:

- Column chromatography
- Ion-exchange chromatography
- Gel-permeation (molecular sieve) chromatography
- Affinity chromatography
- Paper chromatography
- Thin-layer chromatography
- Gas chromatography
- Dye-ligand chromatography
- Hydrophobic interaction chromatography
- Pseudoaffinity chromatography
- High-pressure liquid chromatography (HPLC)
- Super critical fluid chromatography

Table 1.1: The following table summarizes the various phases and techniques in chromatography:

Chromatographic technique	Principle	Stationary phase	Mobile phase
Gas Solid chromatography	Adsorption	Solid adsorbent on a column	Gas
Liquid column chromatography and High-pressure liquid chromatography			Liquid
Thin layer chromatography and paper chromatography		Solid adsorbent on a flat surface	
Gas liquid chromatography and Super critical fluid chromatography	Partition	Liquid on a column	Gas
Liquid column chromatography and High-pressure liquid chromatography			Liquid
Ion exchange	Exchange between the stationary and mobile phase	Solid resin	Liquid
Permeation or Gel exclusion	The materials of the packing material and mobile phase filter out particles	Solid	Liquid

Column chromatography

This is the most ubiquitous mode of operation used especially for the purpose of purification. The stationary phase is packed on a column into which the samples are applied or loaded on the system. The addition of the mobile phase causes flow of the sample and aids in the separation. The use of appropriate systems for support such as fibre glass aids in the flow. Based on the nature of the analytes, fractions are eluted in fashion that is dependent on volume and time (Ozlem Coskun, 2016).

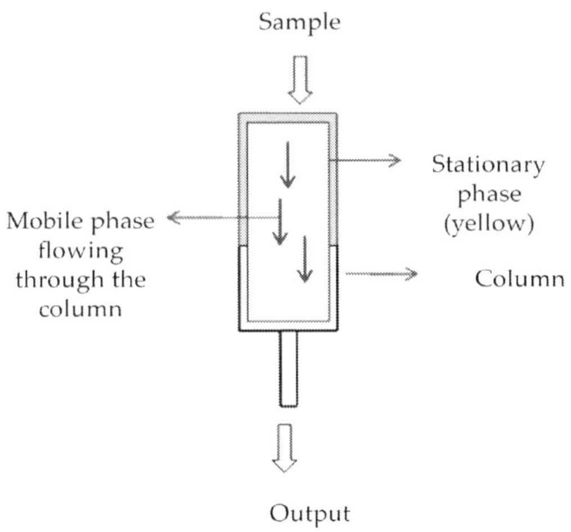

Figure 1.4: Column chromatography

Ion- exchange chromatography

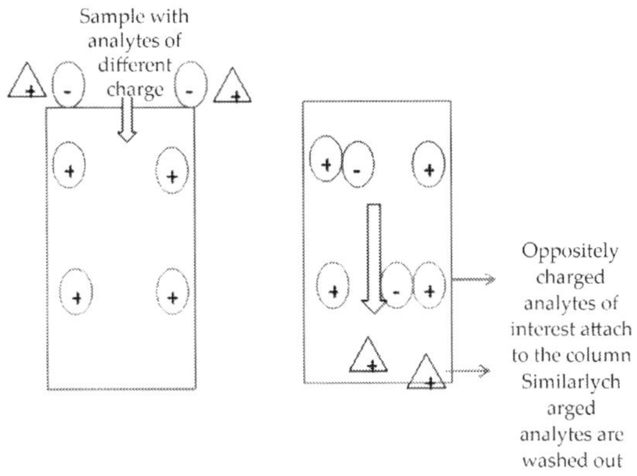

Figure 1.5: Ion exchange Chromatography

This technique uses principles of electrostatic interactions that can develop between charged surfaces of analytes and the matrix in a system. The packing material is designed to possess a charge opposite to that of the charge on an analyte of interest. While matrices that possess negatively charged groups are known as cation-exchange matrices attract analytes with a positive charge, there are anion-exchange matrices that possess positive charges to attract samples bearing a negative charge.

Once bound to the matrix, analytes can be eluted from the column using shifts in charges or pH (11)

Gel- permeation (molecular sieve) chromatography

The technique uses separation on the basis of sizes of samples. The stationary phase particles are chemically non-reactive loaded on a support. When samples are applied on the column or tube, the mobile phase is allowed to flow at a fixed rate. The analytes that are larger traverse through the column without getting into the spaces between the pores. Analytes of small sizes enter the pores and consequently elute out later. Here the stationary phase pores function as sieves to retain certain analytes and allow the passing of others. The major applications of this technique is the separation of particles along with purification and calculation of molecular weight (12).

Affinity chromatography

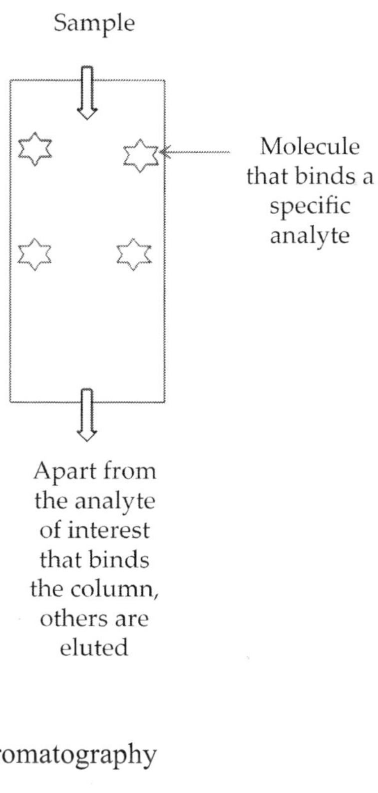

Figure 1.6: Affinity chromatography

This technique uses a ligand that binds to an analyte or sample of interest on a column made of agarose or polyacrylamide. When the samples are applied, the analyte of interest interacts with the ligand and remains in the system while unbound analytes can be washed away. The bound analyte can then be eluted using changes in pH or ionic strength (13). The method is used for isolation several biomolecules such as hormones or amino acids.

Pseudoaffinity chromatography

Using a principle of the above technique, there are reports of the use of azo-dyes or anthraquinones as ligands to capture specific proteins. For instance, immobilized metal affinity chromatography. The technique is used for separation of enzymes such as transferases, dehydrogenases, etc (7).

Paper chromatography

This was one of the earliest systems used that involves a stationary phase made of water molecules in cellulose fibres (filter paper) (14). Following application of samples, the paper is placed in an appropriate solvent that traverses the length of the paper. Based on the solubility of the sample, the analytes of interest are separated.

A term called R_f or relative mobility is calculated as the ratio of the distances travelled by the molecule and the solvent respectively. This qualitative value can aid in the determination of the analytes on the paper (10)

Thin-layer chromatography (TLC)

Using a principle of adsorption, the technique uses a solid adsorbent on a glass plate as the stationary phase. Several materials such as silica, alumina have been reported as suitable materials to be loaded on plates or columns.

The mobile phase is a solvent into which the plate is dipped. The solvent rises up the plate on which a sample is placed due to capillary action. The sample analytes are drawn upwards with the solvent and depending on their properties; the rate of flow decided their separation (15). The identification of samples without colour can employ the use of radioactivity or a chromogenic reaction or florescence to aid in their separation and visualization.

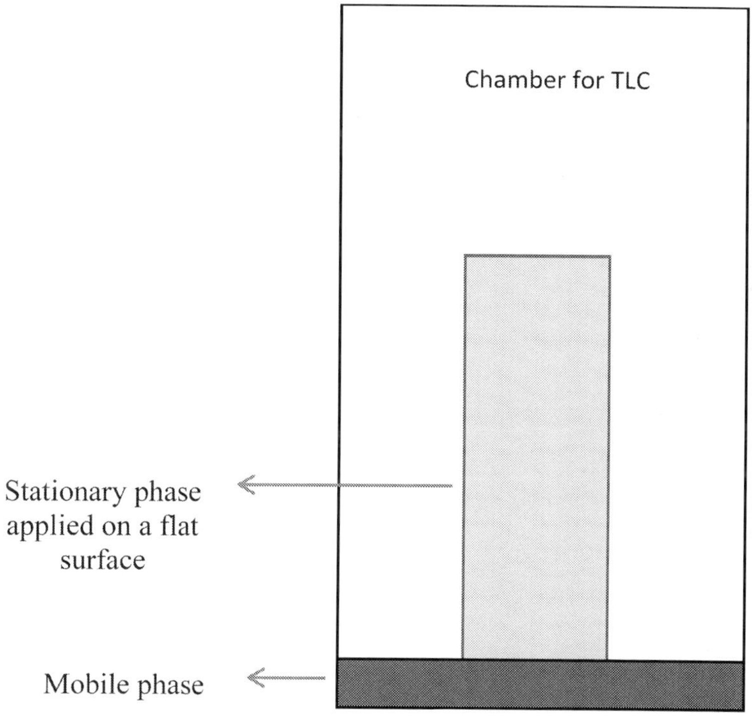

Figure 1.6: Thin Layer Chromatography

Gas chromatography (GC)

The technique was originally designed for separation of volatile fatty acids and has spread to several samples. In the entire instrumentation, is placed a column/tube with a coated with an inert solid with a liquid stationary phase. The mobile phase is a gas: non-reactive such as Helium or Nitrogen or Hydrogen. Following injection of a sample at high temperatures, such volatile samples traverse through the column and are detected. The technique has high sensitivity and rapidity and is favourable for a host of molecules (16).

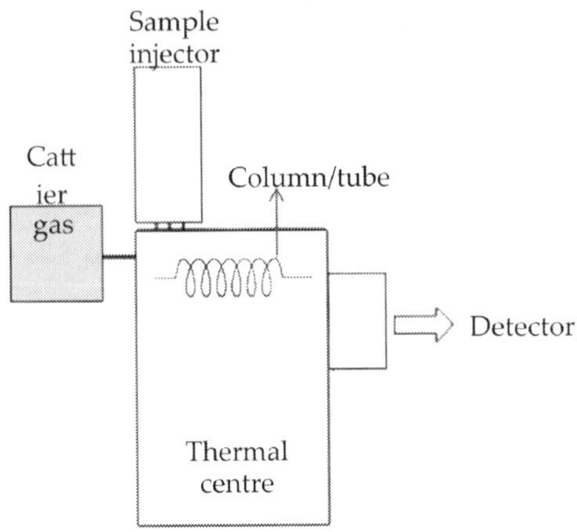

Figure1.7: Gas Chromatography

Dye- ligand chromatography

Based on the ability of several enzymes to bind dyes such as Cibacron Blue F3GA dye that serves as an analog of cofactors such as NAD, the technique uses analogy between ADP-ribose and Cibacron Blue F3GA (9). Other drugs that have been used include: Procion Blue MX-3G or MX-R, Procion Red HE-3B, and Thymol Blue or Phenol Red (17).

Hydrophobic interaction chromatography (HIC)

The technique uses interactions between hydrophobic centres of analytes and the packing materials (18).

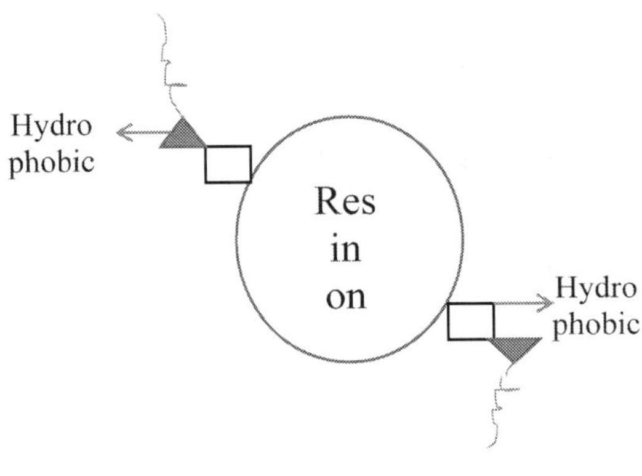

Figure1.8: Hydrophobic interaction chromatography

High-pressure liquid chromatography (HPLC)

The technique uses functional and structural analysis of several analytes in short durations. A mobile phase flows at a high rate (0.1–5 cm/sec) under elevated pressures (up to 400 atmospheres pressure). High resolutions have been achieved with the system using this approach. The instrument comprises a solvent reservoir, a pressure pump, column followed by a detector and data recorder (19).

Super critical flow chromatography (SFC):

The technique was applied for the study of samples of low to moderate masses and extended to chiral moieties. It uses the similar operation of HPLC but instead of a liquid mobile phase, Supercritical fluids are applied.

The super critical fluids possess low viscosity, high diffusion and dissolve many substances. They serve as a mobile carrier phase similar to that of GC yet allow dissolution of samples as in high performance liquid chromatography: hence, the term convergence chromatography is also applied.

A supercritical fluid is the phase of a substance at its critical temperature and critical pressure of the material. Such fluids amalgamate ideal behavior of both gas and liquid phases. Carbon di oxide has been used in its super critical state for such experiments.

SFC present certain advantages such as:

- The use of more varieties of detectors
- Samples that are thermally labile or decompose in GC can be applied.
- Certain analytes pose a challenge in detection in HPLC that can be evaluated in this system.
- The properties can be tuned when a process is running (20).

Flash chromatography:

In a column chromatography experiment, when the mobile phase traverses the column due to percolation or gravity: a term Gravity column chromatography is used. However, the application of pressure to make the mobile phase through the column: the technique is referred to as Flash chromatography. In a column chromatography experiment, following the application of a solvent for elution, the fractions come out slowly. In order to accelerate the speed of the resolution, the application of air pressure has been applied to limit the time of the experiment for a time of 10-15 minutes.

The use of moderate amount of pressure with a short column enables the use of flash chromatography for the discovery of drugs. With the use of smaller particles of silica or alumina or reverse phase silica, as a packing medium and a moderate pressure of 10-15 psi, a separation is quickened: hence the name, Flash (21).

Table 1.2: A glimpse of the highs and lows of few chromatographic techniques

Chromatography	Principle	Summary
Affinity chromatography	The specific absorption based on biological interactions	High resolution and high speed
Ion exchange	Binding of ions based on surface charge	The sample volume is not a constraint apart from high resolution and speed.
Gel filtration	Based on the size and shapes of molecules, they are excluded or included in the pores.	It has optimum resolution and sample volume is a constraint.
Hydrophobic interaction chromatography	The formation of hydrophobic interactions between the groups.	High speed and high resolution reported.

Elution:

This involves the resolution or separation of an analyte from a mixture from stationary phase of the chromatography medium/column. The fractionation of the sample is achieved by elution that can be achieved by several means:

- Shift in pH
- Change in ionic concentration
- Addition of compounds that bind the analyte under study

Elution can be performed in 3 manners: Isocratic, Gradient and Stepwise.

Isocratic elution: The parameters of the elution process are maintained in a constant manner throughout the experiment. A material or eluant is selected that does not form any interactions with the analytes. It is applied when the analytes under study show similar properties.

This elution is simple and does not require elaborate handling or instruments. However, with increased loads, this elution can give rise to the formation of tails on peaks.

Gradient elution: The composition or parameters are subjected to variation to target the isolation of a target analyte. Based on a greater affinity, the analyte is dislodged from the column using varying conditions.

As compared to isocratic elution, a larger range of compounds are separated with reduction in the formation of tails in the peaks. The majority of gradients follow a linear pattern while the experiment can be optimized to achieve other patterns.

Stepwise elution is applied on a regular basis and for large scale experiments. It involves distinct patterns in place of a continually changing gradient (22)

Resolution:

The resolution is defined as distance between the maximum values of the peaks divided by the average of the width of the peak bases. It is a measure of the output of a chromatography experiment and measures both factors: efficiency and selectivity. The former is represented by the formation of peaks that are narrow and show symmetry while the latter is represented as the space between peaks.

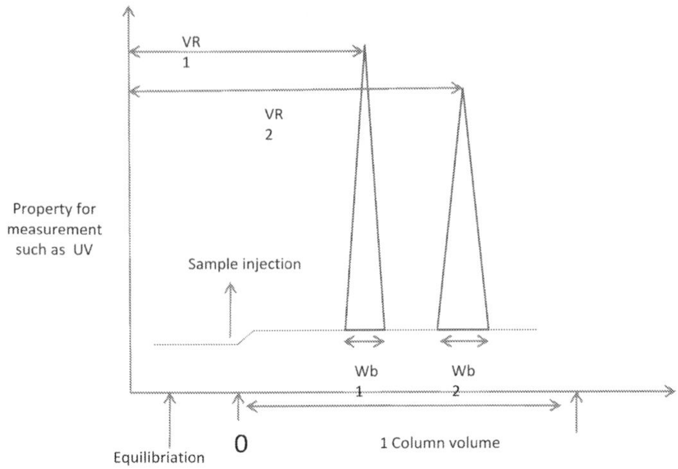

Figure 1.9: Chromatogram: calculation of resolution

Rs can be determined from a chromatogram, as:

Rs= (VR2- VR1)/ (Wb1+Wb2)/2

Where VR2 and VR1 are the peak heights while Wb1and Wb2 are the distances between the peaks

Rs measures the relative separation between two peaks and provides information of the outcome of an experiment. It is recommended that Rs should be greater than 1.5. It is considered that the peak purity is 100% (theoretically 99.9%) (GE Healthcare)

APPLICATIONS OF CHROMATOGRAPHY

The techniques that are based on the *principle of partition* show high efficiency in the separation of several molecules and metabolites. Larger biomolecules such as DNA or RNA can be separated on the techniques of ion exchange or affinity systems.

The separation of proteins can be quickly achieved and also identified using Paper chromatography. The molecular weight studies are facilitated by size exclusion chromatography and Gas chromatography sensitively can resolve several key metabolites.

The technique can be applied in the food industry to assay for the presence of contaminants.

In *medicine*, chromatography yields easy access in clinics and labs with gas chromatography used in the detection of steroids or drugs and pharmaceutically vital molecules. The studies such as kinetics or availability can be evaluated.

In *pharmaceutical research*, the technique aids in the evaluation of new chemical entities to adhere to set standards.

A FEW CASE STUDIES

A few case studies shall be discussed to highlight the contribution of several types of chromatographic techniques:

Affinity chromatography and clinical studies:

The technique of affinity chromatography involves a ligand that is immobilized on a support in order to capture an analyte. An advantage of using this system includes a precision of 1–5% and a time of 5–6 min per sample when applied on an HPLC system. A further advantage is the use of a single ligand in the system can minimize variation between batches of samples applied. A single column can be subjected to several rounds of sample applications using good practices. A few applications of this technique include:

- *The study of displacement of drugs or hormones from proteins.* For instance, The competition for *R*/*S*-warfarin by octanoic acid on HSA; a drug-binding protein in serum. Another study used affinity chromatography to evaluate the displacement of several amino acids such as tryptophan from IISA using bilirubin or caprylate.
- Immunoaffinity columns have emerged as an offshoot of affinity chromatography. For this purpose of chromatographic (or flow-injection) immunoassay, several enzymes are used such as horseradish peroxidase, alkaline phosphatase, and glucose oxidase with labels such as fluorescein, Texas red, or lucifer yellow or fluorescent dye molecule. The binding of theophylline, thyroid-stimulating hormone, transferrin, IgG along with human chorionic gonadotropin have been reported for such assays.

- Chromatographic sandwich immunoassays have found use in the evaluation of intact parathyrin (PTH) in plasma
- Additionally, the extraction of several proteins such as human chorionic gonadotropin in urine or ochratoxin A in human serum, plasma, or milk have been done (Reviewed by David S. Hage, 1999).

Chromatography and Pharma industry:

The technique of chromatography can be applied to several arenas of pharmaceutical research such as:

- New Drug Development
- Purity/Impurity Analysis
- Separations of Isomers
- Support of Biotechnology Products
- Support of Toxicology Studies
- Biopharmaceutic/Pharmacokinetic Studies
- Metabolic Studies
- Clinical Studies
- Forensic Studies
- Diagnostic Studies
- Animal-derived Food Analyses
- Post-mortem Toxicology

The discovery of new drugs necessitates its purity to meet standards set apart from safety from toxicity. As chromatography is an analytical technique, the process can be applied in evaluation or purity of drugs such as the detection of unwanted by products or impurities (23)

Chromatography for Separation of proteins:

The field of proteomics is a valuable tool that requires characterization of proteins. A major tool for this study is the use of Reversed phase high performance liquid chromatography. The technique has been linked with MS detectors to facilitate study with instruments that are the sizes of microchips being reported.

- It was reported that using a Liquid Chromatography–MS/MS method for a breast cancer sample, the quantification of 1600 gene products into 997 protein families was performed in a 2006 study.
- Another study in the same year could isolate successfully a key detoxification enzyme: glutathione S-transferase isoenzyme (GSTs).
- Another technique of Immobilized metal affinity chromatography could achieve the isolation of pure forms of phosphopeptides from Arabidopsis root cell culture.
- The technique of Gel Permeation High Performance Liquid Chromatography or size-exclusion chromatography to separate molecules on the basis of molecular sizes has also contributed to separation. A 2004 study isolated key proteins of rat liver using microcystin-Sepharose chromatography with reports of proteins from peroxisomes.
- Using ligands as the form of Ligand exchange-HPLC involves the use of ion exchange. For instance, major proteins of membrane of Corynebacterium glutamicum were separated and isolated.

Proteomics can use chromatography in the order of: *reversed phase >gel permeation>ligand exchange > affinity chromatography* (24).

LC and Herbal medicines:

The quantitation of many components of Traditional Chinese Medicincs of natural origins has been assisted by the process of LC-QQQ-MS (triple quadrupole mass spectrometer) while the qualitative analysis LC-Q-TOF-MS: quadrupole-time-of-flight ***mass*** spectrometer.

Liquid Chromatography offers an advantage of the separation of complexes and molecules, reviewed by Bingyao Pang et al, 2016 (25).

Chromatography and spurious medicines:

To quote, U.S. Regulation defines counterfeit drugs "as this bought underneath a product call without proper authorization. Counterfeiting can follow to both brand name and generic products, wherein the identity of the supply is intentionally and fraudulently mislabelled in a way that suggests it's far the authentic accepted product. Counterfeit products might also include merchandise without the or with an inadequate amount of the main factor, with the incorrect active factor, or with fake packaging."

The industry of counterfeit medicines stands at a contribution of 7% in the worldwide industry with Southeast Asia, Africa and South America reporting figures of 30%. The field of chromatography has been reported as a useful technique for the detection of such spurious medicines. A few examples are given below:

TLC in combination with UV detection could detect and identify adulteration of phentermine drug. The technique of TLC has identified several important drugs such as ibuprofen, betamethasone, paracetamol and prednisolone, to name a few. In another example of quality check, TLC could detect the purity of drugs against malaria in countries along the Amazon.

The technique of HPLC with UV and diode array detector could aid in the detection of impurities in the bulk and pharmaceutical formulations of sildenafil within 15 minutes.

The technique of HPLC–UV has also been applied to check supplements that are marketed for weight control or reduction. A key contaminant in such drugs is the sibutramine and its analogs. A run in an HPLC column can reveal the presence of several compounds: sibutramine, metformine, diethylpropion, norephedrine, caffeine, modafinil etc. in such formulations.

Another technique is LC along with MS to analyse a molecule but can also aid in the scouring for unknown molecules in complex samples. For unknown samples, LC-MS can be coupled to NMR to characterize novel metabolites or molecules. This is of application for the detection of drugs that have been manufactured by the modification of available drugs.

- LC-MS could aid in the screening of herbal formulations for sibutramine and its analogs. In another study, the presence of adulteration was checked in a host of drugs that included antibiotics, weight loss, aphrodisiacs and psychotropic drugs.
- Gas chromatography (GC) too has been used for testing the illegal drugs apart from its use in checking oils and novel

samples. GC along with MS was applied on Viagra in Hungary to reveal that the pills instead of sildenafil citrate contained amphetamine. Another drug Ritalin was analyzed with GC–flame ionization detection (FID) and GC–MS to report the presence of oxycodone and dihydrocodeinone.

- Reviewed by Eric Deconinck et al, 2013, (26).

Recombinant DNA technology:

With the advent of recombinant DNA technology and the production of proteins for use in therapy and vaccines, purification and isolation of proteins are required. The DNA that is present in such products should have the highest possible criteria due to its potential of integration with genomes of target organisms. This requires that biotechnology products require that no DNA be present at even the picogram-per-dose.

Another application is the separation of DNA and RNA restriction fragments for cloning experiments. It has been reported that ion-exchange columns yield. a resolution of separated nucleic acid fragments: equivalent to that obtained by gel electrophoresis. A 100% recovery rate has been reported for samples of as less as 500 ng using chromatography (23).

Diagnostics:

Gout as well as Lesch-Nyhan syndrome is caused by a lack of xanthine-guanine phosphoribosyltransferase enzyme. An assay using the reaction of the enzyme with substrate, hypoxanthine to yield the product, inosine-5'-monophosphoric acid can be applied on HPLC.

The use of chromatography aids in the quantification of both substrate and enzyme to assay the activity of enzyme in an accurate manner that in turn reflects on the presence or absence of the syndrome or disease (23).

CONCLUSION

Chromatography has completed more than a century in the field of research. It initially started off with a separation of color and then extended to several fields. The technique is reported as sensitive as well as efficient with several types including Column chromatography, Ion-exchange chromatography, Gel-permeation (molecular sieve) chromatography, Affinity chromatography, Paper chromatography, Thin-layer chromatography, Gas chromatography, Dye-ligand chromatography, Hydrophobic interaction chromatography, Pseudoaffinity chromatography, High-pressure liquid chromatography ,Super critical fluid chromatography and Flash chromatography. Each of these techniques have distinct applications as well as advantages and challenges.

The process of chromatography has wide applications ranging from the analysis and separation of samples to drug discovery and quality checks. It can also be applied in the diagnosis as well as the characterization of new compounds.

REFERENCES

1. Ettre LS. Nomenclature for chromatography. Commissions On Chromatography And Analytical Nomenclature. IUPAC Recommendations. 1993.
2. Ettre. MS. Tswett and the Invention of Chromatography. LC-GC Europe .September 2003: 2-7.
3. Tswett M Ber. Dtsch. Botan. Ges. 1906; 24, :316–323.
4. Berezkin VG, Ed., *Chromatographic Adsorption Analysis: Selected Works of M.S. Tswett*. 1990. Ellis Horwood, New York, USA 1990.
5. Lederer E. J. Chromatogr. 1972;73: 361–366.
6. Curling J. Process Chromatography: Five Decades of Innovation. Supplement to Biopharm International. 2007;10-19.
7. Porath J. From gel filtration to adsorptive size exclusion. J Protein Chem. 1997;16:463–8.
8. Harris DC. Exploring chemical analysis. 3rd ed. WH. Freeman & Co; 2004.
9. Coskun O. Separation techniques: Chromatography. North Clin Istanb. 2016;3(2): 156–160.
10. Donald PL, Lampman GM, Kritz GS, Engel RG. Introduction to organic laboratory techniques. 4th ed. Thomson Brooks/Cole. 2006;797–817.

11. Amercham Biosciences. Ion Exchange chromatohgraphy, Principles and methods, Amercham Pharmacia. Biotech SE. 2002;751.
12. Helmut D. Gel Chromatography, gel filtration, gel permeation, molecular sieves: a laboratory hand book. Springer-Verlag; 1969.
13. Firer MA. Efficient elution of functional proteins in affinity chromatography. J Biochem Biophys Methods. 2001; 49:433–42.
14. Porath J. Immobilized metal ion affinity chromatography. Protein Expr Purif. 1992; 3:263–81.
15. Sherman J, Fried B, Dekker M. New York, NY: Handbook of Thin-Layer Chromatography; 1991.
16. Kouremenos KA, Johansson M, Marriott PJ. Advances in Gas Chromatographic Methods for the Identification of Biomarkers in Cancer. Journal of Cancer. 2012; 3:404-420. doi:10.7150/jca.4956.
17. Hage DS. Affinity Chromatography: A Review of Clinical Applications. Clin Chem. 1999;45(5):593–615.
18. Queiroz JA, Tomaz CT, Cabral JM. Hydrophobic interaction chromatography of proteins. J Biotechnol. 2001; 87:143–59.
19. Regnier FE. High-performance liquid chromatography of biopolimers. Science. 1983:245–252.
20. Gopaliya P, Kamble PR, Kamble R, Chauhan CS. A Review Article on Supercritical Fluid Chromatography. Int J Pharm Res Rev. May 2014;3(5):59-66.
21. Avare P, Khanvilkar V, Chalak N. Flash chromatography: Area and applications. Pharma Tutor. 2014;2(5): 89-103.
22. GE Healthcare. Strategies for Protein Purification Handbook. 12 28-9833-31 AA.
23. Ahuja S. Chromatography and Pharmaceutical Analysis. Am Chem Soc. 1992: 0097-6156/92/051: 1-6.
24. Ali I, Aboul-Enein HY, Singh P, Singh R, Sharma B. Separation of biological proteins by liquid chromatography. Saudi Pharm J. 2010; 18: 59–73.
25. Pang B, Zhu Y, Lu L, Gu F, Chen H. The Applications and Features of Liquid Chromatography-Mass Spectrometry in the Analysis of Traditional Chinese Medicine. Evid-Based Complementary

Altern Med. 2016; 2016, Article ID 3837270, 7 pages. doi:10.1155/2016/3837270

26. Deconinck E, Sacré P-Y, Courselle P, De Beer JO. Chromatography in the Detection and Characterization of Illegal Pharmaceutical Preparations. J Chromatogr Sci. 2013; 51 (8): 791-806. doi: 10.1093/chromsci/bmt006.

CHAPTER 2

COLUMN CHROMATOGRAPHY

INTRODUCTION

Currently with the help of chromatography over 60% of chemical analysis is done worldwide (1). It comprises of a mobile phase and a stationary phase. While the mobile phase travels through the stationary phase gathering up the compounds to be tested, it takes the compounds with it. Results of any chromatography are held within the stationary phase as different components of the compound are wrapped up, movement is stopped and the components are separated along with the mobile phase.

Chromatography depends on the degree to which the solutes between two immiscible phases tends to combine the stationary phase that is attached bed of large surface area, while the other is termed as the mobile phase in which fluid travels through the surface of the mixed phase.

Column Chromatography is a routine preparative technique in which a mixture of molecules is separated using different sizes of column and is used to purify compounds depending on their polarity or hydrophobicity on scales from micrograms up to kilograms. The classical preparative chromatography column is a glass tube with a diameter from 5- 50 mm and a height of 5 cm -1 m with a tap and at the bottom a filter of a glass frit or glass wool plug to prevent the loss of the stationary phase. Two methods the dry method and the wet method which usually requires mesh 70 – 230 (63 – 200 µm) silica gel are commonly used to prepare a column (1).

COLUMNS

Columns available in the department are simple glass tubes, to control the solvent flow commonly have a stopcock attached, and to support the adsorbent may have a fritted plate, to act as a solvent reservoir some have a flask attached to the top of the column and it differs in size which allows one to choose the best column for the separation (1, 2). The larger diameter column can handle larger quantities of compound. Resolution increases with increasing length, and decreases with increasing diameter of the adsorbent in the column. Therefore, better separation will be provided by 25 g of adsorbent in a column of 1 cm diameter than in a 2-cm column.

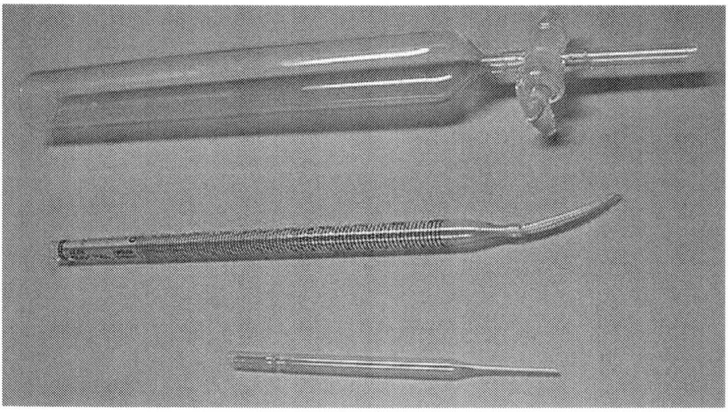

Figure 2.1: Different types of columns used for Column chromatography

Column chromatography contains of two different phases: the solid phase and the mobile phase. To setup as the solid phase the column generally Silica or Alumina is used. In order to set up the column, slurry consisting of the adsorbent along with a non-polar solvent needs to be stirred well to eliminate all the air bubbles so that sample can run evenly in the column (2,3). The prepared slurry with sand at the top to help in loading the sample is added to the column.

PREPARATION OF THE MICRO COLUMN

Using a Pasteur pipette the micro column can be prepared. To avoid the silica from escaping the column, glass wool is put at the bottom, which is flat to ensure an equal separation. Silica is then added to the column and to avoid uneven separation a metal spatula is used to pat the column so that the top of it is also level. In order to compact the silica the column should be washed with a non-polar solvent. Once it has been wetted with the solvent, the solvent is added continuously to it to avoid its drying. Once it dries up, the separation will be uneven leading to incorrect results.

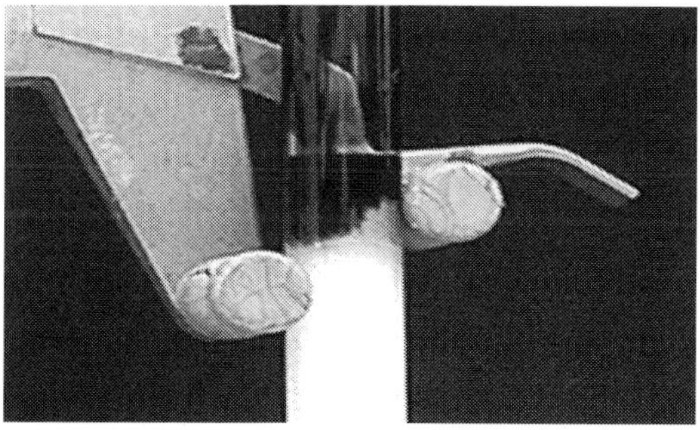

Figure 2.2: Sampling

The crude mixture uniformly added at the top of the column as the solvent level has just about reached to the same height as the sand (3). To get good separations the band is made as thin as possible and while adding the crude mixture, any movement of the solid silica medium is avoided. Additional solvent can be added to elute when the extract reaches the silica.

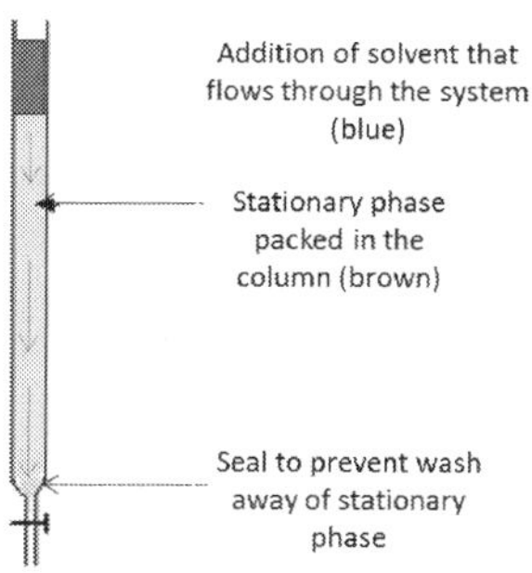

Figure 2.3: The schematic of a typical column used for chromatography

COLUMN BACKING

The adsorbent is applied to the Column in two ways

Slurry packing (Wet method): To avoid air bubbles the adsorbent is suspended in the mobile phase and mixed well, then poured into the column. Before the slurry application, at the tap end of the column a piece of glass wool or cotton must be added. Sand may be added after the slurry (3).

The column must be allowed to settle overnight. In gel chromatography to captivate the mobile phase and swell the adsorbent must be saturated in the mobile phase overnight.

Dry Packing: The dry adsorbent is transferred to the column directly. To avoid air bubbles vibration is used then the mobile phase as travel through the adsorbent. For this method Gel Chromatography cannot be functional.

MOBILE PHASE

It is a mixture of organic solvents (unusually one solvent only) which is generally selected after TLC studies in variety solvent systems. For decent result of all materials to be separated by the column Rf value less than 0.6 must be made by solvent system (4). Separation will not happen if the system movement is more and produces higher Rf and if they do not move spots at all on TLC are not good for column separation.

Isocratic system: In this process, separation is accomplished using the same mobile phase from the start to the finish.

Gradient system: during separation by increasing the proportion of the more polar solvent the polarity of the system increases. A typical gradient may be start with CHC13, followed by CHC13/MeOH mixtures with gradual increase in % of MeOH till all spots are eluted from the system.

ADDING THE SAMPLE

Only a minimum amount of solvent is used when adding the sample to the column. After the packing is done the sample can be filled directly to the top of the column. Usually, a minimum amount (5-10 drops) of a polar solvent is used to dissolve the mixture (1, 4). Without disturbing the flat top surface of the column, the solution is carefully added using a pipette to the top of the column. When adding solvent eluent, the top level could be disturbed. To keep the top of the column level after the sample is loaded, a small layer of white sand is added to the top. The solvent eluent is continuously added while collecting small fractions at the bottom of the column after the mixture is added. Through this process, ensure that the protective layer of sand is in place, with a pipette as sand reduces disorder of the column and thinning the sample.

Collecting small fractions (1-3ml) is important to the success of the column separation and very small fractions can be pooled together whereas if the collected fractions are too large, more than one compound can be acquired. Chromatography is to be repeated for better separation in such situation.

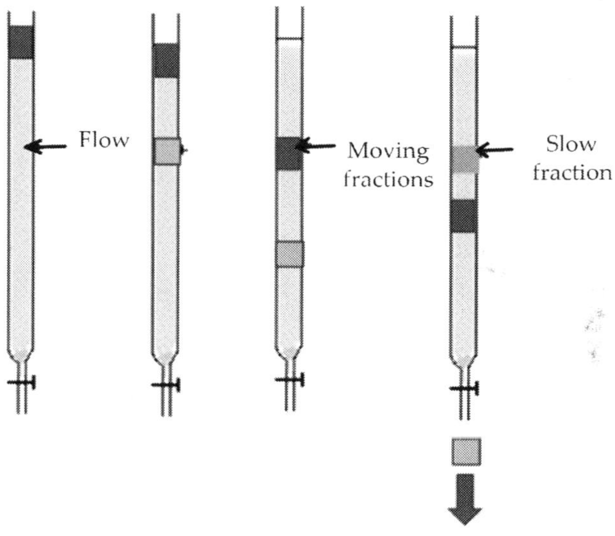

Figure 2.4: Fraction collection

The collected fraction is to be labelled. While collecting the elutes, the test tubes are changed more often to that the closely associated bands do not get mixed up coming out of the column.

FLASH COLUMN CHROMATOGRAPHY:

Flash can help speed up the flow rate of the column:

Column chromatography is very time taking process as the solvent is allowed to elute through the column one drop at time. Flash Chromatography is a method to speed up the process, and uses a pressure of about 10 psi of air or nitrogen to force the mobile phase through the column. In general, this process does not give a good separation because of the rate of the mobile phase is increased. But, it can be used to raise the speed without lowering the quality of the separation by using a finer grade of alumina or silica (5).

Silica gel grades in the technique is mesh 230 – 400 (40 – 63 μm). To make sure rational flow rates, the particle size of the solid was kept larger than150 to 200μm; even then, flow rates were at best a few tenths of a millilitre per minute (6). Because increase in flow rates are accompanied with increase in plate heights and decrease in column efficiency, attempts to speed up this classic procedure by application of vacuum or pressure were not effective. Until the late 1960s, the technology for producing and packing with particle diameters as small as 5 to 10 μm was developed, although earlier it was realized that large decreases in plate heights could be expected to accompany decreases in the particle size of packing. The name high-performance liquid chromatography (HPLC) is often employed to differentiate these newer procedures from their predecessors, which still find great use for preparative purposes.

Automated systems

Components like a gradient pump, sample injection ports, a UV detector and a fraction collector, normally found on HPLC systems to gather the eluent. Usually, multiple injections on prep-HPLC mechanised systems offer sample separation that are much cheaper and quicker and offer a few milligrams to many kilograms on industrial scale.

Flash chromatography is classified into two types

- *LPLC* - Low pressure liquid chromatography system which operate around 50 -75 psi.

- *MPLC*: Medium pressure liquid chromatography systems which operate above 150 psi (4)

The software controlling an automated system organize the components, let a user to only gather the groups of target compound (supposing they are evident on the system's detector) and comfort the user to find the bring about purified material in the fraction collector.

Procedure for Running of flash column

The column packing

To fill to stopcock hole in the chromatography column, it is plugged with a small piece of cotton wool. To make the diameter of the sand about the same as the column, 2 cm sand is added. It is recommended to add dry silica and in most cases, it is best if it is about 6 to 10 inches. About 1-2 cm of sand is added to the top of the column. The solvent is transferred with the vacuum still applied, (premixed, i.e. 4:1 hexanes/ethyl acetate) and allowed to flow though the column till it is more or less eluting. To ensure complete packing enough solvent is in the column for 5-6 column volumes worth to flow though. At this instant, the stopcock is closed and remove the vacuum line is removed; keeping in mind not to let the column run dry, all of the solvent is eluted with air pressure and when the solvent level is parallel the process is stopped by means of the sand avoiding any cracks or patches. The solvent eluding from the stopcock is avoided to be warm or hot (7).

Column Loading:

A solution of the reaction or compound mixture, is added carefully to the top of the silica with a pipette. The solvent level is allowed to run below the sand into the very top of the silica gel after each accumulation. 2-3 pipettes of chromatography solvent is added cautiously and this is pushed into the column. The left-over space is filled with compressed air, preferably about 2 inches/minute flow rate. When a reaction mixture or compound is not soluble in a suitable solvent for loading, the compound is dissolved in acetone, adding silica gel and concentrating the gel to dryness. The dry silica is then added to the top of the packed silica column.

Running the column:

The 13 mm test tubes for small scale (i.e. 5-50 mg) and larger test tubes for bigger columns fractions are collected according suitable for the type of column and polarity. After adding the compound, the fractions are instantly collected as it doesn't take time for very non-polar compounds to elute from the column. Poor parting and reduced yields is due to slow flow of the compounds on the silica gel. The fractions are checked to see which ones contain the desired compounds, followed by spotting of each fraction or so on a TLC plate. Portions with the same compounds are mixed, the test tubes carry away with methylene chloride or distilled ethyl acetate, and the solvent concentrated under reduced pressure (8)

Column cleaning:

Smooth air over the column for ~2 hours will give dry, free flowing silica gel. The contents of the column are transferred into the silica waste container and the column is washed with water and acetone, or hint of liquid soap. Try to avoid scratching the columns with abrasive brushes or soaps.

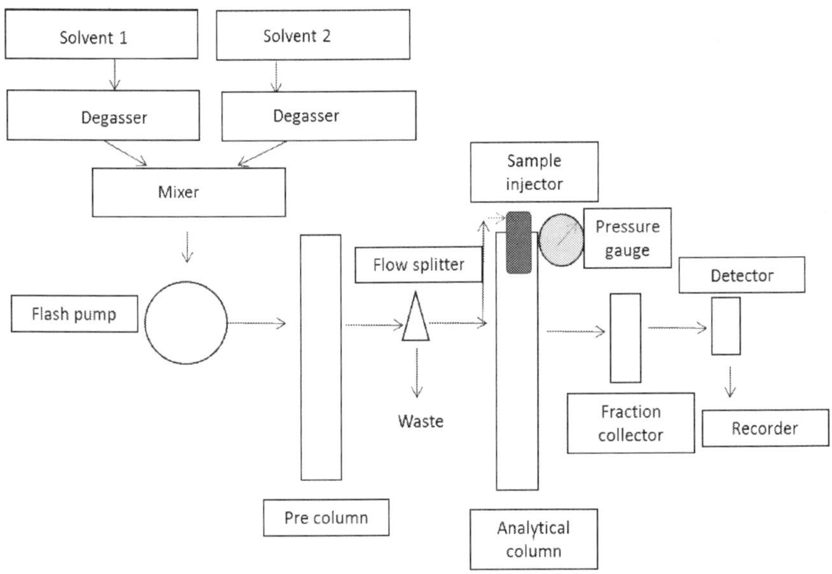

Figure 2.5: Schematic of Flash chromatography

ADVANTAGES

Flash chromatography has several advantageous:

It can be used for the synthesis processes in laboratory, being a fast and economic method,

Best for the separation of compounds up to gram quantities.

No costly equipment necessary.

The automated changes between normal phase and reversed phase chromatography is convenient

DISADVANTAGES

It is a time-consuming technique the amount of solvents required are high, which makes the process expensive

LOW AND MEDIUM PRESSURE COLUMNS:

By using pumps that generate low or medium pressure the movement of mobile phase is accelerated. The increase in the flow rate cuts the time of separation (7).

VACUUM COLUMNS (VLC)

The adsorbent is applied dry into a sintered glass funnel by dry method or as solution. Then to gather each fraction the mobile phase is added portion by portion and after each portion vacuum is functional.

A FEW TERMINOLOGIES AND CALCULATION OF MIGRATION

Differential migration is different compounds move through the system with different rates of movement. By the number of molecules of that compound in the mobile phase the speed of any compound in the mixture is determined (11)

U: velocity of solvent (mobile phase). z Ux: velocity of material X. z R: fraction of material X in the mobile phase. Ux = UR

If R = I ie all the compound molecules are present in the mobile phase.

Ux = U X I Ux = u

Matenial X will move with the solvent velocity.

z if R = 0.0 ie all the compound molecules are existing in the stationary phase.

Ux = u x 0.0 :. Ux = 0.0

Material X will not move at all. For any material to be separated through the system it must be distributed between the mobile phase and stationary phase.

Capacity Factor, z: The factor that control the distribution of any material between the two competitive phases is called "Distribution coefficient" or "Capacity factor" or "Mass distribution ratio" K' (n)s K' = (n)m

(n)s: Is the total number of moles of the compound in the stationary phase.

(n)m: Is the total number of moles of the compound in the mobile phase.

K' = (tr – t0)/ t0

tr: Time required for the sample to cross the column (retention time).

t0: Time required for the solvent molecules to cross the column

a higher K' means that the material retained more time on the column ei move slowly. Smaller K' means faster movement.

MOVEMENT OF MATERIALS THROUGH THE CHROMATOGRAPHIC SYSTEM IN THE FORM OF "ZONES" OR "BANDS"

It is assumed that the chromatographic system is composed of number of "distribution systems" or "equilibrations" called "Theoretical Plates". Each of it is composed of stationary phase and mobile phase. The height of each plate is called "*Height equivalent to Theoretical Plate*" (HETP). The number of theoretic plates "N" is vital for parting. Increasing "N" caused in finer bands and well separation.

N = L/HETP;

L: Column length; “N” can be increased by Increasing the length of the column (impractical), and decreasing the HETP.

The HETP can be decreased by decreasing the particle size of the stationary phase and Proper selection of good mobile phase.

FACTORS AFFECTING SEPARATION

Factors due to Stationary Phase:

Particle size: Decreasing the particle size increases the surface area and separation improves. Yet, decline of the particle size will drop the flow rate of the mobile phase.

To achieve very good separation in HPLC, *very fine particles* are used. By the help of high pressure pumps to drive the mobile phase over the stationary phase the flow rate problem is solved. To withstand the high pressure columns are prepared of stainless steel.

Adsorbent activity: The selection of suitable adsorbent is vital.

Uniformity in column packing: bands will be uneven if the stationary phase is not packed uniformly and causing in poor separation.

Concentration of the mixture: the size of stationary phases and the apt ratio separation between samples is very important as too much samples lead to in bad separation (4, 5, 7)

Factors due to Mobile Phase:

A proper mobile phase: An extremely polar mobile phase will wash out all components without any separation. On the other hand, very non-polar mobile phase will result in broad bands and poor separation.

Rate of flow: Slower flow rate usually results in a better separation and narrower bands.

Consistency of flow: Endless flow of the mobile phase in the whole experiment provides good separation than disturbing the flow (4, 5, 8)

Factors due to Columns:

Column dimensions: Increasing the length of the column improves separation but hints to slower flow rate and at times is not practical.

Column temperature: Increasing the temperature generally reduces the adsorption power of the stationary phase and increase elution speed, which may lead to decrease in the competence separation.

DIFFERENT TYPES OF COLUMN CHROMATOGRAPHY

Four different types of column chromatography which works on four different mechanisms are as follows:

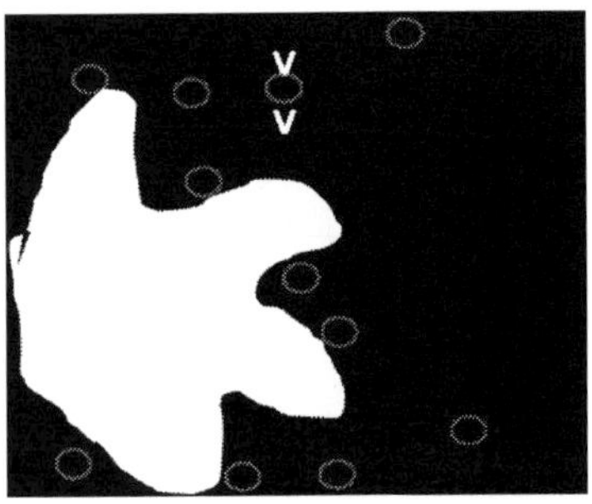

Figure 2.6: The Principle of Adsorption Chromatography

Adsorption chromatography: the components of mixture are selectively adsorbed on the surface of packing column material, i.e. adsorbant (Figure 2.6) (9).

The stationary phase of this technique is a solid material on which the sample compounds are adsorbed and mobile phase is either a liquid (solid-liquid chromatography) or a gas (gas-solid chromatography). Adsorption is completely different from absorption. Adsorption chromatography is based on the interface between the solute molecules and on the go sites on the stationary phase. In this technique, if the stationary phase is more polar than the mobile phase, the high polar compounds in the mixture will tightly bind to the stationary phase while less polar compounds will informally bind to the stationary phase. Less firmly bound compounds will be eluted out by the mobile phase earlier than the tightly bonded ones.

Partition chromatography: the component is partitioned between the mobile phase and their stationary phase held stationary on passive solid support (9).

The stationary phase of this chromatography technique is a liquid supported on an inert solid. Mobile phase is either a liquid (liquid-liquid chromatography) or gas (gas-liquid chromatography). This is based on the difference in solubility is also due the differences in polarity. More polar substances have higher affinity to the stationary phase. Less polar substances will go into the mobile phase (Figure 2.7).

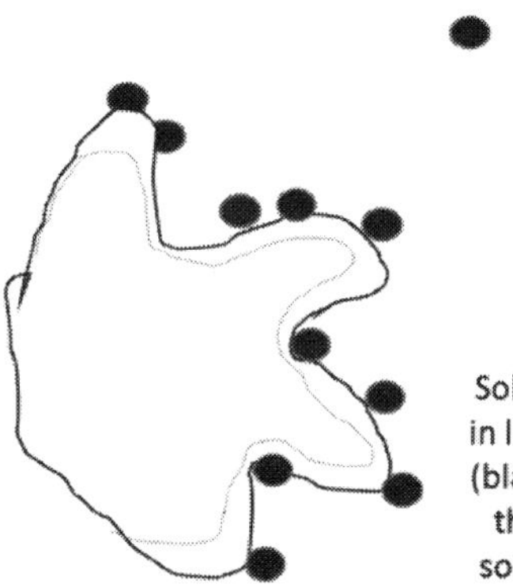

Figure 2.7: The principle of partition chromatography

Ion exchange chromatography: the constituent of sample is selectively retained by exchange resin by replacing ion/s on packing material. The stationary phase of this technique is an ion-exchange resin but the mobile phase is a buffered aqueous solution. The resin can be either cationic or anionic. Ion exchange columns are cation exchange columns (contains negatively charged groups) and Anion exchange columns (contain positively charged groups). In here the charged analytes are involved to the counter ions (exchange groups) in the column. Analyte ions are eluted out using a buffer with a different pH to weaken the electrostatic interactions between the analytes and the exchangers. Analytes are present in the mobile phase whereas the exchange groups present in the column. The details of this technique are discussed in chapter 10 of this book.

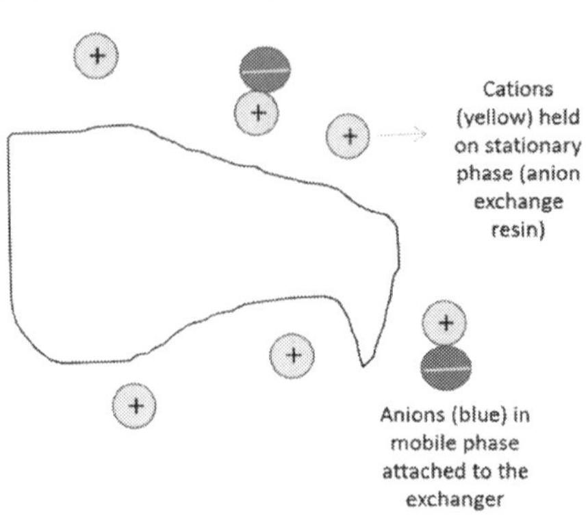

Figure 2.8: The principle of ion-exchange chromatography

Gel chromatography is the method in which column is packed with a permeable gel which brings out separation by putting through a sieve or molecular filtration action.

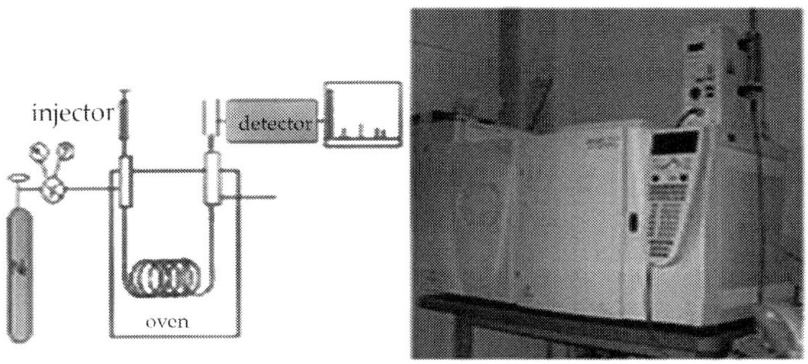

Figure 2.9: The principle of gel chromatography and its instrument

Gas chromatography is a type of adsorption chromatography. Because the mobile phase of GC is a gas, this is labeled as "gas chromatography". The stationary phase can be a solid/liquid adsorbent.

If the stationary phase is a solid adsorbent, the technique is called as "gas-solid chromatography" where as if it is a liquid adsorbent, it is called as "gas-liquid chromatography" (non-volatile liquids are used). GC helps in many ways to detect explosive air pollutants, essential oils and very useful forensic studies. The GC will be discussed in detail in subsequently (chapter 5).

APPLICATIONS:

- used for Purification of Protected Peptide (12)
- used for Separation of Closely Related Organic Compounds (13,14)
- For purification of trace compounds from organic mixtures Flash systems are powerful tools.
- tool to monitor the reaction progress and to isolate and identify a mixture's compounds

High Speed Flash Fractionation of Natural Products – Tocopherols

As the preliminary isolation step, reversed phase flash chromatography allows the tocopherols to be concentrated and have fewer oil contaminants thereby increasing the lifetime of the HPLC columns (14)

Refining Natural Product Purity by Orthogonal FLASH Purification (14, 17)

In this application, several solvent systems were evaluated by TLC.

Cleanse, gather and classify the various avermectin components in a semi-synthetic extract and also the isomers of an aqueous modified antibiotic precursor. The goal for this work was to isolate each isomer with > 98% purity (11, 12, 13, 14)

Many pharmaceutical companies use The Flex system to cleanse compounds in the process of drug discovery (15)

Amino modified silica is used with normal-phase solvents and because the surface chemistry is slightly alkaline is apt for nitrogen heterocyclic purification.

As a lead investigation and optimization tool in drug discovery it has received increased attention.

Case study 1

Separation plant pigments Carotenoids & Chlorophylls ***(18, 19).***

Biological materials containing carotenoids and their solutions must be protected from the action of light. For the **isolation** of plant pigments from biometabolites mixtures, the physical and chemical properties are taken into account as tey are sensitive to heat, light, oxygen, alkaline bases, and acids. The exposure to direct light causes cis–trans photoisomerization, and subsequently to their photodestruction. Many carotenoids are thermolabile (xanthophylls); their heating being indicated only when it is absolutely necessary.

The separation of carotenoids is done at room temperature or up to –20°C, in the dark. In the case of *hot saponification* they should be protected by a low–boiling solvent (30–60°C). Carotenoids are sensitive to oxygen in the adsorbed state (in thin layer or column chromatograms) and may be oxidized in the presence of oxygen or peroxides, thus making it necessary to operate in inert conditions (under nitrogen or vacuum). The nitrogenous atmosphere minimizes the oxidation during the extraction and saponification. When the carotenoids are exposed to acids, changes such as oxidative decomposition, cis–trans isomerization and isomerization of 5,6–epoxides and 5,8–epoxides occur. However, neutralization by (calcium carbonate, pyridine, dimetilalanine) minimizes the inconveniences. Chlorophylline, astaxanthene, and carotene represents typical groups of plant pigments. The sample (0.5mL) was injected in a 75 x 40mm C18 flash cartridge that was pre-wetted with water. The green chlorphylline band was eluted with ~120mL of 60% methanol in water, and astaxanthene was eluted with ~140mL of 100% methanol, while the carotene band stayed essentially at the origin. Spectral methods have the advantage over the chemical methods of identification, being faster, more accurate, require smaller amounts of material and enable continuous analysis at different stages of processing enabling its recovery without altering the composition of the bio substance under investigation.

Case study 2

Purification of diatom pigments

Fucoxanthin, diadinoxanthin and diatoxanthin are carotenoids found in brown algae and most other heterokonts, involved in photosynthetic and photoprotective reactions, and they have numerous possible health aids. They can be extracted from diatom *Phaeodactylum tricornutum* by sonication, extraction with chloroform: methanol and preparative thin layer chromatography. An additional column chromatography step was used in purification of these pigments which increased the purity of fucoxanthin and endorsed for concentration of diadinoxanthin and diatoxanthin before HPLC separation. The enhanced protocol is useful for gaining high purity pigments for biochemical studies (16).

The separation of carotenoids from Chls was achieved with additional column chromatography step, an important step, as free Chl is a photosensitizer and can, upon illumination, lead to the generation of reactive oxygen species (20, 21). This in turn may contribute to a lower pigment quality and purification yield. Additional column chromatography step decreases the time when carotenoids are mixed with Chls and hence lowers the chance of carotenoids being damaged by the photodynamic effect caused by free Chls. The separations where petroleum ether is used are completed slightly faster than those where the extraction gasoline is used (20-23). This may be attributed to the lower density of petroleum ether, as it consists of shorter hydrocarbon chains.

Spectral quality of purified pigments: The column separation step did not cause a significant improvement in respect to spectral quality (general shape, peak positions and ratios), no matter which solvent was used. Furthermore, no isomerization was detected by this analysis. In some spectra, minuscule peaks originating from Chl were present (not shown) more frequently when no column pre-separation step was employed. This supports the idea that additional column chromatography step increases the purity of the isolated pigments, while the spectral quality of the main constituent remains comparable.

Purification procedure yield: Pre-separation with column chromatography did not lead to the increase of the purification yield. It can be attributed to the fact that the additional purification procedure causes greater material losses. On the other hand, column pre-separation step eliminates the Chls from the extracts and allows for the separation of more concentrated samples on HP-TLC.

Pigment purity: The highest pigment purities obtained were above 93%. In the case of all tested pigments, the pigment purity (assessed by HPLC analysis) did not differ significantly between the two solvents used in the column chromatography step.

Further purification using HPLC: To remove contaminations, the samples of were subjected to further purification, using HPLC (the same method as described above). The results demonstrated a purity exceeding 99.5% with room for improvement. In addition, a shorter version of the protocol for was developed for preparative HPLC separation, which comprises of the same solvents and takes only 20 minutes (16).

The additional pre-separation step, utilizing the column chromatography, significantly enhances the purity and lowers variability. The column pre-separation step is an easy and cheap way to concentrate the samples before HPLC separation, which lowers the overall cost of the isolation procedure. While this step decreases the overall yield, it can limit the potential photodynamic effect from Chl, as the carotenoids and Chls are separated earlier in the process (23). Thus, adding a column pre-separation step can be useful for small-scale laboratory preparations and it allows the inexpensive concentration of pigments before further purification and analysis. The improved purification protocol can be used for obtaining material for further studies, concerning the biochemical and biophysical properties promote the advances in research focused on these pigments.

REFERENCES

1. Kondeti RR, Mulpuri KS, Meruga B. Advancements in column chromatography: A review. World J Pharm Sci. 2014; 2(9): 1375-1383
2. Column/Flash Chromatography as a Sample Preparation Technique, In Sample Preparation
3. Fundamentals For Chromatography. Agilent technologies. http://www.dioksyny.pl/wp-content/uploads/Sample-preparation-wg-AGILENT-2014.pdf
4. Fair JD, Kormos CM. Flash column chromatograms estimated from thin-layer chromatography data. J Chromatography A. 2008; 1211(1–2): 49-54.
5. Ayare P, Khanvilkar V, Chalak N. Flash Chromatography: Area & Applications. PharmaTutor. 2014; 2(5):89-103.
6. Danot M, Nahmias S, Zoller U. An undergraduate column chromatography experiment. J. Chem. Educ. 1984; 61 (11), p 1019. DOI: 10.1021/ed061p1019

7. William CS, Hill DC. General methods for flash chromatography using disposable column. Mol. Divers, 13(2), 2009, 247-252.
8. Chaudhari H, Chaudhari F, Patel M, Pradhan PK, Upadhyay UM. A Review on A Flash Chromatography. International Journal of Pharmaceutical Development & Technology. 2012; 2(2):80-84.
9. Hansen SH, PerHelboe, UllaLund. Chapter 4.1. Types of liquid column chromatography. New Comprehensive Biochemistry. Volume 8, 1984, Pages 151-153
10. Introduction to Modern Liquid Chromatography, Second ed; snyder LR and kirkland JJ. A Wiley-Interscience Publication JOHN WILEY & SONS, INC. 1079
11. Chromatography in Modern Chemical Techniques, Royal Society of Chemistry.
12. Kamysz W, Okrój M, Łempicka E, Ossowski T, Łukasiak J. Fast and Efficient Purification of Synthetic Peptides By Solid-Phase Extraction. Acta Chromatographica, NO. 14, 2004
13. Dane GD, Raka KC, Honde BS, Bhawal GS, Tajane PJ. Review on Flash Chromatography. Int J Pharm Rev Research. 2013; 3(1):|45-49.
14. Swathi G, A. Srividya, A. Ajitha, Dr. V. Uma Maheswar Rao. Review On: Flash Chromatography. World J Pharm Pharm Sci. 2015;4(08): 281-296.
15. Friesen JB, McAlpine JB, Chen S-N, Pauli GF. Countercurrent Separation of Natural Products: An Update. Journal of Natural Products. 2015;78(7):1765-1796. doi:10.1021/np501065h.
16. Tokarek W, Listwan S, Pagacz J, Leśniak P, Latowski D. Column chromatography as a useful step in purification of diatom pigments. Acta Biochim Pol. 2016;63(3):443-7.
17. Roge AB, Firke SN, Kawade RM, Sarje SK, Vadvalkar SM. Brief Review On: Flash Chromatography. Int J Pahrm Sci Res. 2011; 2(8): 1930-1937.
18. Zhao S, Tuan PA, Kim JK, Park WT, Kim YB, et al. Molecular characterization of carotenoid biosynthetic genes and carotenoid accumulation in Lyciumchinense. Molecules. 2014;19: 11250-11262.

19. Butnariu M (2016) Methods of Analysis (Extraction, Separation, Identification and Quantification) of Carotenoids from Natural Products. J Ecosys Ecograph 6: 193. doi:10.4172/2157-7625.1000193
20. Marder JB, Droppa M, Casp V, Raskin VI, Horváth G. Light independent thermoluminescence from thylakoids of greening barley leaves. Evidence for involvement of oxygen radicals and free chlorophyll. Physiologia Plantarum. 1998; 104: 713–719. http://dx.doi. org/10.1034/j.1399-3054.1998.1040428.x
21. Krieger-Liszkay A. Singlet oxygen production in photosynthesis. J Exp Bot. 2005;56: 337–346. http://dx.doi.org/10.1093/jxb/erh237.
22. Kraay GW, Zapata M, Veldhuis MJW (1992) Separation of chlorophylls c1, c2, and c3 of marine phytoplankton by reversed phase-C18-high-performance liquid chromatography. J Phycol 28: 708–712. http://dx.doi.org/10.1111/j.0022-3646.1992.00708.x.
23. Ritchie RJ (2006) Consistent sets of spectrophotometric chlorophyll equations for acetone, methanol and ethanol solvents. Photosynth Res 89: 27–41. http://dx.doi.org/10.1007/s11120-006-9065-9.

CHAPTER 3

PAPER CHROMATOGRAPHY

INTRODUCTION

Paper Chromatography is an analytical technique which allows the separation of complex mixtures into its simpler and individual components by using two immiscible phases. The separation of complex is essential in any chemical analysis. Hence the compound needs to be simplified to its constituents. Scientists can reveal the exact components present in mixtures like cigarette smoke, proteins, plant enzymes, etc. (1). It is a simple, inexpensive, quick and a powerful technique which requires minimum sample for analysis. Important studies like understanding the mechanism of photosynthesis in plants or inventions like synthesizing proteins like insulin and many more would be difficult without the knowledge of paper chromatography (2).

HISTORY

Mikhail Tswett's studies on plant pigments, particularly the carotenoids and chlorophyll led to the invention of chromatography in 1901. His experiment included a vertical glass column consisting of an adsorptive matter like calcium carbonate powder. He added plant pigment solutions on top of these columns with organic solvents. The pigments attached to the powder resulting in the formation of bands of various colors on the column. He named the technique as Chromatography (chroma-color graphein-writing) (3) as it separated components in a series of colors. Adsorption chromatography was also suggested by Tswett for the analysis of colorless compounds. Richard Willstätter was studying chlorophyll and he introduced Chromatography to the Western world. Using Chromatography techniques, Willstätter discovered two major types of chlorophyll: blue-green and yellow-green type. Further he discovered a similarity between hemoglobin and chlorophyll also finding the difference of an iron and magnesium atom in the structures respectively. This finding led to an increase in the crop yield using his previous data and hence he was honored with the Nobel Prize for Chemistry in 1915 for his commendable work in chemistry (4)

Paper chromatography was then extensively used resulting in many important inventions. Frederick Sanger in 1918 used it to study the insulin molecule finding the number of amino acids and their order (5).

PRINCIPLE

Paper Chromatography works on the principle of separation of complex compounds into its individual components by the use of a liquid mobile phase and a stationary solid phase. The stationary phase is a uniform absorbent paper and the liquid used to cause the separation of the mixture is a chromatographic solvent, wash liquid, or developer. A sample mixture (solute) is applied on a piece of filter paper (stationary phase) with its edge immersed in the solvent (liquid phase).

The solvent moves upwards on the paper by the capillary movement of liquids, carrying the components of the given substance along its path. The mixture contains products having different degrees of attraction to both- the stationary phase and the mobile phase. When the mobile phase is allowed to move along a stationary phase with the sample of mixture on it, the components of the mixture that have more affinity to the stationary phase move slowly. The components of the mixture having less affinity to the stationary phase move upwards quickly with the mobile phase. The stationary phase is allowed to dry and is sprayed with a reagent that shows a change in color and it also reveals distinct bands of components, which can be further analyzed.

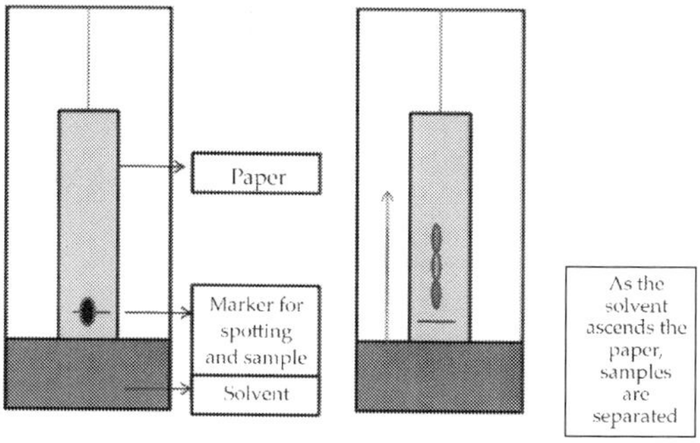

Figure 3.1: Principle of paper chromatography

WORKING

A drop of sample is placed near the end of the porous paper strip and allowed to dry. The strip is immersed a bit into the solvent not immersing the sample. The solvent ascends along the paper, timed appropriately to suit the experimental setup.

The paper acts as an adsorbent and it has strong affinity to polar molecules held by Van der Waals forces and hydrogen bonds. The strip is then allowed to dry in an air tight container to avoid evaporation from the surface of the paper. The paper thus shows different bands of the separated sample components depending on polarity or hydrogen bonds (6). More color bands on the paper indicate more components of the sample. Different chemical components have different structures and hence each molecule will have different polarity thus giving each molecule a different solubility to the given solvent. Thus, various color bands are formed by the various components of the given sample depending upon its affinity and polarity with the paper. In-case the solvent is not able to separate the sample components, the paper may be turned 90°, repeating the process with another solvent.

NATURE OF THE PAPER

The paper used is made up of cellulose which is a polar material. It is made of homopolysaccharide of glucose which is highly purified. If the components of the sample are polar, they form bonds with the cellulose paper and don't travel far, whereas if the components of the sample are non-polar, they travel farther. Recently, modified papers impregnated with ion exchange resins, silica gel, alumina, etc are available and mostly, Whatmann filter papers are used. Composition of Whatmann filter paper no.1 is: 98-99% cellulose, 0.015-0.1/% Ether soluble matter, 0.4-0.8% Pentosans and 0.07-0.1% Ash (7).

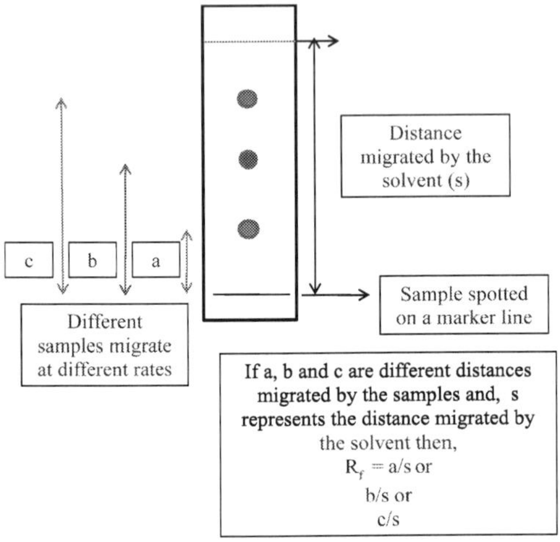

Figure 3.2: Calculation of R_f paper chromatography

RETENTION FACTOR (R_F):

The identification of the separated components of a given sample is done by calculating the retention factor (R_f). Distance covered by one component is constant relative to the solvent provided factors like composition of the solvent and the type of paper are kept constant. It also depends on the solvent used in an experiment and its temperature (8,9).

$\mathbf{R_f}$ = Distance travelled by the solute/distance travelled by the solvent

TYPES OF PAPER CHROMATOGRAPHY

Components separate according to their solubility. Different components have various characteristics and hence paper chromatography has five development techniques. This differs depending on the formation of the individual bands either along gravity or against gravity.

Ascending Paper Chromatography:

In this technique, the spot of sample is placed near the solvent and the chromatogram is immersed a bit in the solvent (not touching the sample) which is in the bottom of the apparatus. The solvent rises against gravity and is hence termed as Ascending Paper Chromatography. The solvent carries the components of the sample along with it. The least polar component will be at the end of the chromatogram while the most polar component will be at the top of the tank. This is a comparatively slow process.

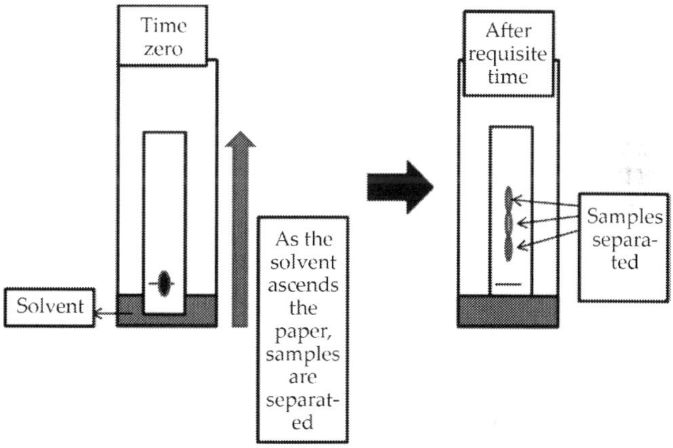

Figure 3.3: Ascending Paper Chromatography

Descending Paper Chromatography:

This technique has a complex apparatus setup. The chromatogram is attached to a support and the mobile phase is filled in the chamber. The chromatogram is saturated with stationary phase before being hung. The solvent carries the sample components in the downward direction, along gravity. Here, the least polar component will be found at the bottom of the strip whereas the most polar component will form a band at the top of the strip. Since in this technique the flow of components is along gravity, it is a quick process compared to ascending paper chromatography.

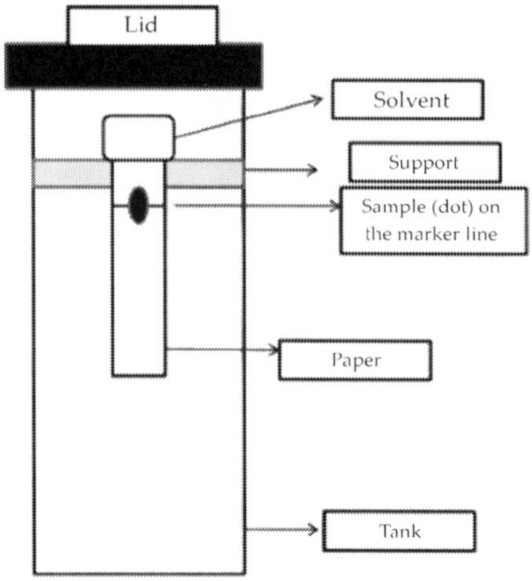

Figure 3.4: Descending Paper Chromatography

Ascending-Descending Paper Chromatography:

This technique is a hybrid of ascending and descending paper chromatographies. The upper region of the ascending chromatography is folded over a rod. Later the paper is descended after the rod is crossed (10)

Radial Paper Chromatography:

In radial paper chromatography, the sample spot is placed in the centre of a circular paper. The paper is tied horizontally parallel to the Petri dish containing the solvent. A wick of paper is dipped in the solvent which rises through the wick by the capillary movement of liquids resulting in the separation of components in concentric circles (11).

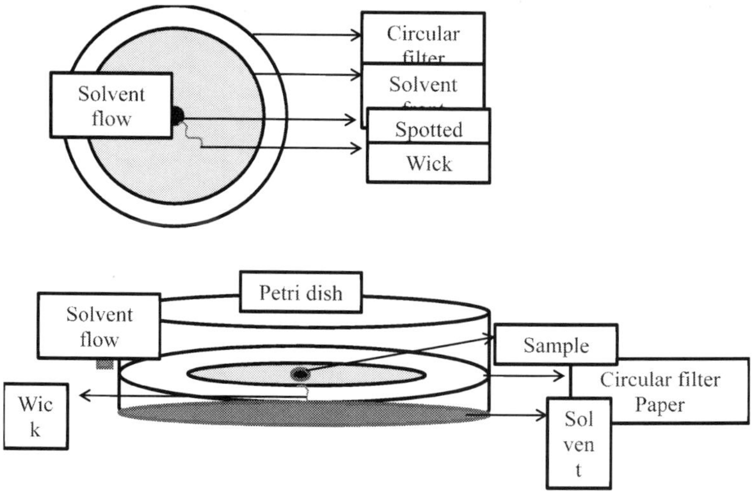

Figure 3.5: Radial Paper Chromatography: Side view **(below)** and Top view **(above)**

Two-Dimensional Paper Chromatography:

A rectangular or square paper is designed for the ascending paper chromatography and after a few hours it is turned 90° clockwise in the direction of the first run using a different solvent (12).

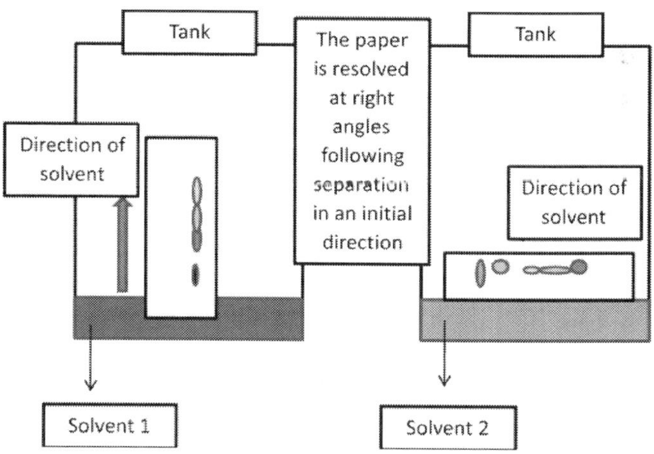

Figure 3.6: 2D Chromatography

CASE STUDY 1

In 1962, G.B. Crump performed experiments to separate mixtures of phenols using paper chromatography technique. This technique was found applicable to a huge range of petroleum products. The phenols were coupled with diazotized aromatic amines forming stable azo dyes which were separated to its individual components using formamide impregnated paper. This series of experiments were successful for the identification and separation of phenols in wide range of products (13).

CASE STUDY 2

Melvin Calvin worked with green algae cells to find the path through which solar energy was converted to carbohydrates. Green algae cells were introduced with radioactively labeled Carbon to see the location of its action in the cells. He used two-dimensional paper chromatography for the same which led to the separation of compounds present in the green algae. Calvin identified more than ten intermediate products that were created during his experiments successfully by using paper chromatography (14).

CASE STUDY 3

Scientists used paper chromatography to analyze samples of human plasma phosphatides. Human plasma phospatides were separated quantitatively and qualitatively by paper chromatography. A component having properties of inositol phosphatide, sphingomyelin and lysolecithin phosphatidylethanolamine were derived from the separation technique. Limitations and advantages of the technique for plasma phosphatides analysis were further discussed (15).

CASE STUDY 4

Vischer and Chargaff performed experiments to separate and quantitatively estimate the amounts of purines and pyrimidines using paper chromatography. The technique helped them in the identification of the separated minute samples of pyrimidines and purines (16).

APPLICATIONS OF PAPER CHROMATOGRAPHY

- For the estimation of free amino acids in the tissues of rats (17)
- For the detection of contaminants in foods and drinks
- For the study of ripening and fermentation
- To check the purity of the pharmaceuticals
- For the detection of drugs in living beings.
- For the detection of adulterants in a given sample
- For the detailed analysis of cosmetics
- For the analysis of the reaction of mixtures in biochemical labs. Eg: For the analysis of Riboflavin, its derivatives and nucleotides (18)
- For the separation of phenolic estrogens in human urine (19)
- For the detection of steroids evaluating many compounds (20)
- Paper chromatography was also used to analyze the individual components of various classes of compounds like purines, pyrimidines, acidic p-dimethylaminobenzaldehyde reacting compounds, organic acids, amino acids and many more (8).

ADVANTAGES OF PAPER CHROMATOGRAPHY

- It is the cheapest and simplest compared to other chromatography methods.
- It works with very little sample size.
- Being a simple apparatus, it occupies less space compared to those of other techniques.
- Unknown organic and inorganic components can be identified.

DISADVANTAGES OF PAPER CHROMATOGRAPHY

- It is not effective in quantitative analysis.
- Ascending paper chromatography is a very slow technique.
- Complex and large sample size can't be applied and separated.
- It is less accurate compared to HPTLC or HPLC.

- Dipping the sample dot in the solvent accidently, results in repetition of the procedure.

CONCLUSION

The technique that employed a piece of filter paper dipped in solvent and used the separation of compounds on the basis of mobility came as a revolution in the field of science. Though initially the technique faced a lot of opposition, it subsequently emerges as a leading tool in the field of technology. Initially students keen on science are exposed first to this technique to estimate biomolecules such as proteins. Using a small amount of sample and minimum apparatus, the technique achieves separation in a rapid manner.

REFERENCES

1. Harborne JB. The chromatographic identification of anthocyanin pigments, Journal of Chromatography. 1958; 1:473-488.
2. Fenton EL. A method for the assay of insulin by paper chromatography. Biochem J. Mar 1959; 71(3): 507-513
3. Tswett M. Physikalisch-chemische Studien **über das Chlorophyll. Die** Adsorptionen (Physical-chemical studies of chlorophyll. Adsorption.). Berichte der Deutschen botanischen Gesellschaft. 1906; 24:316–323.
4. Synge RL. Tsvet, Willstatter, and the use of adsorption for purification of proteins. Arch Biochem Biophys. 1962 Sep;Suppl 1:1-6.
5. Stretton AOW. The First Sequence: Fred Sanger and Insulin. Anecdotal, Historical and Critical Commentaries on Genetics Edited by James F. Crow and William F. Dove. Genetics. 2002; 162: 527–532.
6. Smith I. Chromatographic and Electrophoretic Techniques, Chromatography, 3rd edition. Great Britain at the Pitman Press, Bath. William Heinemann Medical Books Ltd.; 1969; 1:1.
7. Richard J. Block Raymond Le Strange Gunter Zweig. Paper Chromatography. 1st Edition; 1952; Academic Press.
8. Fink K, Cline RE, Fink RM. Paper chromatography of several classes of compounds: correlated Rf Values in a variety of Solvent

Sytems. Anal Chem. 1963; 35(3): 389-398.

9. Hais IM. Factors which influence paper chromatographic RF values. J Chromatograhy A. 1968; 33:25-27.
10. McCullough T, Rocabado W. Combined ascending-descending paper chromatography. *J Chem Educ.* 1992;69 (12): 995.
11. Hendrickson MJ, Berueffy RR, McIntyre AR. Radial Paper Chromatography...Application to Analysis of Mixtures of Chlorophyll Derivatives. *Anal Chem.* 1957;29(12): 1810–1815.
12. Baugh P. Advances and Changes in the Techniques of Multi-Dimensional and Comprehensive Chromatography and When Coupled with Mass Spectrometry. J Anal Bioanal Tech. 2016;7: 4 323.
13. Crump GB. The paper chromatographic separation and identification of simple phenols. J Chromatography. 1962;10: 21-28.
14. Benson AA, Bassham JA, Calvin M, Goodale TC, Haas VA, Stepka W. The Path of Carbon in Photosynthesis. V. Paper Chromatography and Radioautography of the Products. *J Am Chem Soc.*1950;72(4): 1710-1718.
15. Marinetti GV, Albrecht M, Ford T, Stotz E. Analysis of human plasma phosphatides by paper chromatography. Biochimica et Biophysica Acta. 1959;36(1): 4-13.
16. Vischer E, Chargaff E. The separation and quantitative estimation of purines and pyrimidines in minute amounts. J Biol Chem. 1948;176: 703-714.
17. Awafaba J. Application of paper chromatography to the estimation of free amino acids in tissues. Arch Biochem. 1948; 19: 172-173.
18. Cbammer JL. Paper Chromatography of Flavine Nucleotides. Nature. 1948; 161: 349-350.
19. Dao T. A method for the separation of four phenolic estrogens in human urine by paper chromatography and an evaluation of biological assay of these compounds. Endocrinology. 1957;61(3): 242-255.
20. Kritchevsky D, Kirk MR. Detection of steroids in paper chromatography. Arch Biochem Bioph.1952; 35(2):346-351

CHAPTER 4

THIN LAYER CHROMATOGRAPHY (TLC) ANDHIGH PERFORMANCE THIN LAYER CHROMATOGRAPHY (HTLC)

INTRODUCTION

Thin-Layer Chromatography (TLC) is a simple, flexible and cost-efficient separation technique for both qualitative and quantitative analysis, enabling simultaneous analysis of many substances with minimal time requirement. Hence, an account of its rapidity and simplicity the technique can be applied in several platforms. There are appropriate apparatus that are applied for each level of the instrumentation (1,2).

The technique of thin layer chromatography show similarity with that of LC systems, the instrumentation involved is more simple and economical. The technique also enables the separation of several samples all together, for instance: up to sixty samples in one run.

PRINCIPLE

As compared to LC, the technique of TLC yields lesser resolving powers yet the technique is more economical. The applications of TLC still show an upward trend in spite of developments of LC.

TLC shows a limitation in its application over multiple samples as the plate has a limited amount of samples that can be applied. The technique uses samples applied on a plate while the technique of LC involves elution from samples applied on a column (1,2).

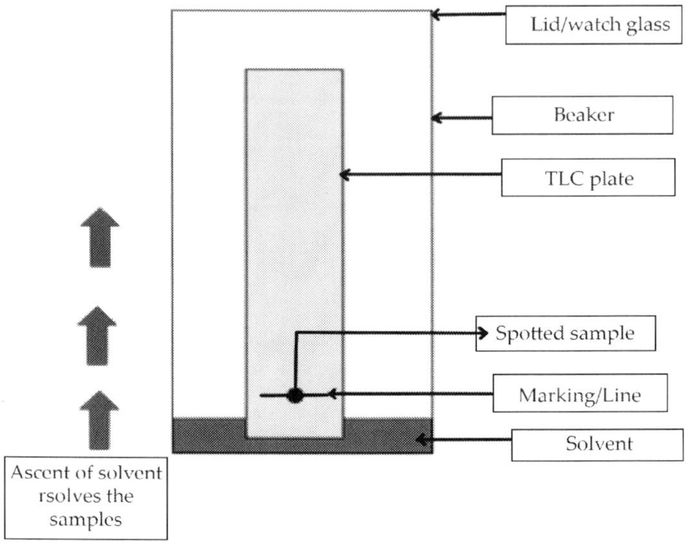

Figure 4.1: Thin Layer Chromatography Chambers

Solvent must be prevented from evaporating from the plate surface during development as this would cause the composition of thedeveloping solvent to change.In the simplest case, the developing chamber can consist of a round or square glass jar fitted with a glass cap. Sufficient solvent mixture is placed in the chamber to raise the level about a centimetre from the base. The plate is then placed in the chamber with the end where the samples have been placed dipping into the solvent and the cap replaced. The sample spots must not be at, or below, the surface of the solvent mixture or they will be washed from the plate. To improve the air space saturation with solvent, the walls, or part of the walls, of the chamber are sometimes covered with filter paper to act as a wick that soaks up the solvent and provides a greater surface area for evaporation (2).

The saturated solvent vapor in the chamber not only prevents solvent evaporating from the plate surface but partly controls the retention mechanism by surface deactivation. The solvents are selectively adsorbed on the surface of the stationary phase causing the solutes to interact, not with the native silica surface, but with the silica surface covered with the most strongly interacting solvent.It should be emphasized, however, that the equilibrium between the *solvent vapor* and the plate will not be the same as the equilibrium between the *solvent* and the plate. For example, for a binary mixture of solvents having concentrations of solvent in the gas phase of c_1 and c_2 respectively,

$$c_1 \neq X_1 \ and \ c_2 \neq X_2$$

where (x_1) is the molar fraction of solvent (1)

and (x_2) is the molar fraction of solvent (2)

In fact, *Raoult›s Law* applies and

$$\frac{c_1}{c_2} = \frac{X_1 P_1}{(1 - X_1) P_2} \ where \ X_2 = (1 - X_1)$$

and (P_1) and (P_2) are vapor pressures of pure solvents (1) and (2) respectively. A typical set up of the experiment is as follows:

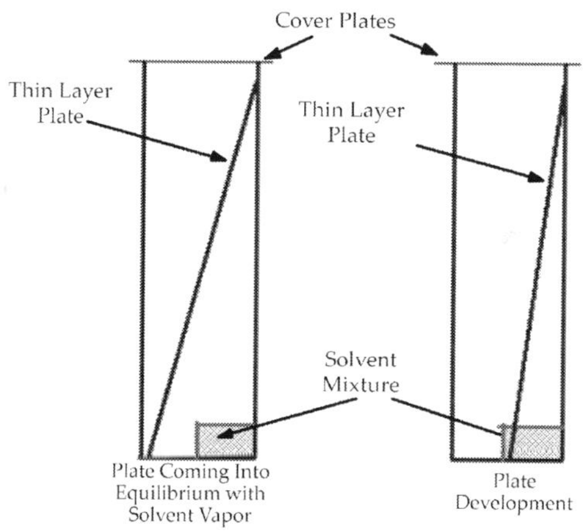

Figure 4.2: Dual chamber TLC

The apparatus shows similarity with that of that applied for the regular method but differs in the presence of the solvent in a different container.

The plate is placed in the enclosure and allowed to return into equilibrium with the solvent vapour for a few minutes. The plate is then lifted and placed so that the end now dips into the developing solvent to continue with the process as standardized. Two outcomes are plausible:

- There is an increase in solvent velocity as compared to that of the plate that is not saturated.
- The silica surface of the plate that is not saturated adsorbs a considerable amount of solvent at the front

The advantage conferred by the movement of the front on the presaturated plate in a rapid way is balanced by the presence of the spots that are very close to the solvent front on the unsaturated plate. Though these effects may be insignificant on the resolution of samples, the impact can play a big role in the case of certain mixes.

CONTINUOUS PLATE DEVELOPMENT

Due to small surface areas offered by thin layer plates, the use of novel apparatus has aided the procedure of development.

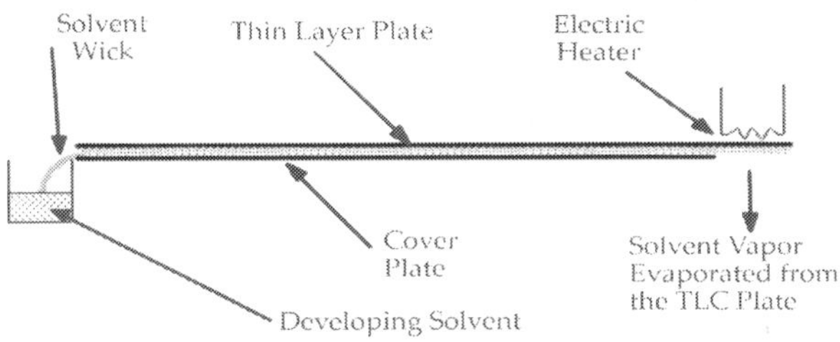

Figure 4.3: Continuous Plate Development: Thin Layer Plate

The plate for chromatography is placed such that the stationary phase faces a descending direction. It is placed on another plate. The solvent is transferred from a reservoir between the two plates to the stationary phase. The solvent traverses through the column by virtue of surface tension. In order to circumvent evaporation of the solvent, the apparatus is placed within a container. This is followed by an application of heat to the periphery of the plate to facilitate solvent evaporation. It has been reported that the process resembles and assumes the formation of a LC system (2,3).

SAMPLE APPLICATION

The application of samples must involve the spotting as a spot of small dimensions. In a regular system, up to 2ml of samples are spotted as a standard practice. The usage of high performance TLC plates (HPTLC plates) necessitates the sample to be applied in volumes of nano proportions: 100-200 nl as the particles on the plate themselves are 2.5-3.5mm in radius. An HPTLC plate requires that the sample is applied in the shape of a circle and the dimensions specified are below 1mm.

This entails the use of volumes that are of nano proportions that require concentration. The use of a micro-pipette or micro-syringes can aid in sample application followed by evaporation of solvent. The process is subject to repetition till the sample volume is satisfactory.

The concentration of sample can be effected by the use of mild heating in the vicinity or caution. The use of syringes that are manipulated by computational methods can add an element of automation to facilitate the application of focused samples with small dimensions.

The device consists of a plate with a small indentation, at the center of which is an aperture that can be connected either to a vacuum or to a source of gas pressure. A polymer film is placed over the plate surface and a vacuum applied to suck the film into the indentation. A small quantity of sample solution is placed in the indentation and the solvent evaporated. This procedure is repeated until sufficient sample is present on the film for a satisfactory TLC separation (1).

It is important that the sample is ***not*** evaporated to ***dryness*** as the transfer of solid materials to the thin layer plate can be ***very inefficient***. When sufficient sample has been accumulated and the droplet is still liquid, the thin layer plate is placed over the film and a positive pressure now applied to the aperture at the center of the indentation.

The film extends to the surface of the TLC plate and the sample is transferred by contact, as a result of surface tension forces and the adsorptivity of the TLC plate surface. The plate is then conveyed to the solvent chamber and the separation developed in the usual way.

The width of the ***concentrating zone*** is about 2 cm consisting of big silica particles coated to occupy a low area that causes a lower retention.

The ***concentrating band*** is coated closely adjacent to the normal retentive coating which, consists of the usual particles 5-7 mm in diameter but with a much higher surface area and, thus, much greater retentive capacity.

Multiple samples can be applied on the concentrating zone in volumes of microliter dimensions in a tandem or sequential format followed by solvent evaporation to allow for concentration. This allows the development of the sample in the form of a broad band that traverses and occupies the concentration zone. The samples undergo retardation due to increased retention. The technique presents an edge over others: Contaminants such as salts or polymers get trapped in the zone of concentration and the possibility of them affecting the resolution of the separation.

Band applicators operate differently and are usually fully automated. The sample is atomized in a stream of air or nitrogen depending on the nature of the sample and its tendency to oxidation.

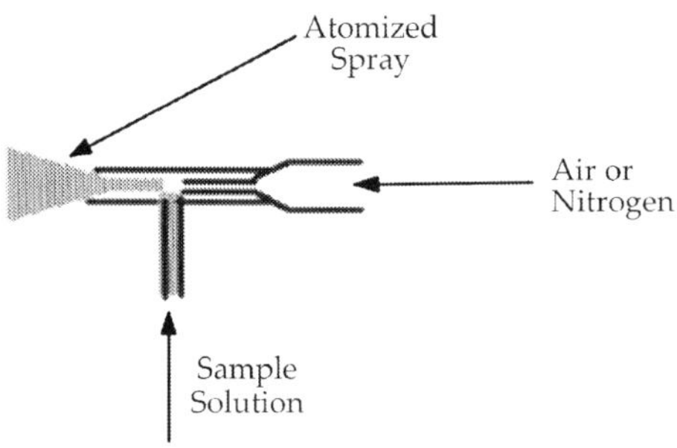

Figure 4.4: The atomizer is mounted on a mechanical arm that can sweep from side to side while directing the atomized sample on to the surface of a TLC plate.

The range and the number of sweeps, are usually under computer control and the speed of movement is adjusted such that the solvent is able to evaporate from a given area of sample before it receives the subsequent dose. After dosing, the plate is developed in the normal way.

Some exceedingly novel and clever devices have been developed for TLC. These devices indeed improve the performance of the TLC analysis but are also expensive and in many cases tend to make the TLC system more like a liquid chromatograph. *The great advantage of TLC is its low cost and its relatively high separating capability*. If the required performance required is at the limit or beyond the capability of the technique, there is no point in trying to stretch it. The rational solution for the chemist or analyst would be tochange to an alternative procedure such as liquid chromatography or to some other technique if more appropriate (1,2).

APPLICATIONS OF THIN LAYER CHROMATOGRAPHY (TLC):

- *Purity of a compound*: TLC can facilitate the checking of purity of a desired substance. The contaminants or impurities are seen in the form of additional spots on a plate when the substance is run with a pure standard.
- *Establishing identity of compounds:* The identification of several molecules such as oils or lipids can be performed using TLC apart from the resolution.
- *Studying reactions:* The status of progress of a reaction can be facilitates through the use of TLC.
- *Biochemical analysis:* TLC is extremely useful in isolation or separation of biochemical metabolites or constituent from its body fluids, blood plasma, serum, urine etc.
- Pharma field: The detection of impurities has relied on the method of TLC. Additionally, the resolution of drugs with several ingredients can be separated. The testing of quality has been done for several compounds that include pain killers or steroids or tranquilizers etc.
- Chemical applications: The resolution and study of compounds or ions with similar characters can be achieved through the use of TLC.
- Food and cosmetics: The studies of artificial preservatives or coloring agents can be achieved by TLC.

HIGH-PERFORMANCE THIN-LAYER CHROMATOGRAPHY (HPTLC)

Touted ahead of other techniques, High-Performance Thin-Layer Chromatography (HPTLC) uses several layers of chromatography in conjunction with latest technology to affect the various aspects entailed in the protocol:

- Application of samples in accurately.
- Developing the chromatogram effectively.
- Evaluation using appropriate softwares.

HPTLC meets all quality requirements of today's analytical labs, even in a fully regulated environment (3).

The expenditure involved in the technique of HPTLC is far smaller than other techniques with lower costs applied for a sample. The visualization of the samples as they resolve on the plate is a hallmark of the technology. The use of Information Technology enables the development of images.

Table 4.1: Comparison between TLC and HPTLC

Criteria	TLC	HPTLC
Sorbent layer thickness	250μm	100μm
Spotting of the sample	Manual	Auto sampler
Sample volume	1-5 μL	0.1-0.5 μL
Solid support	Silica gel, Kiesul-guhr, Alumina	A range of stationary phases like silica gel for normal phase; C8, C18 for reversed phase modes
Mobile Phase	More amount	Require comparatively less amount of mobile phase
Analysis Time	Slow (20-30 min)	Faster because shorter migration distance (3-20 min)
Scanning	Not possible	Qualitative and quantitative scanning of the entire chromatogram using UV/ Visible/ Fluorescence scanner that is also a type of advanced densitometer
Efficiency	Less	Higher, due to smaller particle size generated

COMMON METHODOLOGY FOR HPTLC

Analysis Method development in thin-layer (planar) chromatography is one of the most significant steps for a qualitative and quantitative analysis. During establishing a new analytical procedure, always starts with wide literature survey i.e. primary information about the physicochemical characteristics of sample and nature of the sample (structure, polarity, volatility, stability and solubility). It involves considerable trial and error procedures (4). During method development, stationary phase selection should be based on the type of compounds to be separated. General steps involved in HPTLC method developments are as follow:

HPTLC uses smaller plates (10*10 or 10*20 cm) with significantly decreased development distance (typically 6 cm) and analysis time (7–20 min). The technique offers several advantages such as:

- Increase in resolution.
- Enhanced detection.
- A boost in the quantitation of samples.

MOBILE PHASE SELECTION AND OPTIMIZATION

The selection of mobile phase is based on adsorbent material used as stationary phase and physical and chemical properties of analyte.

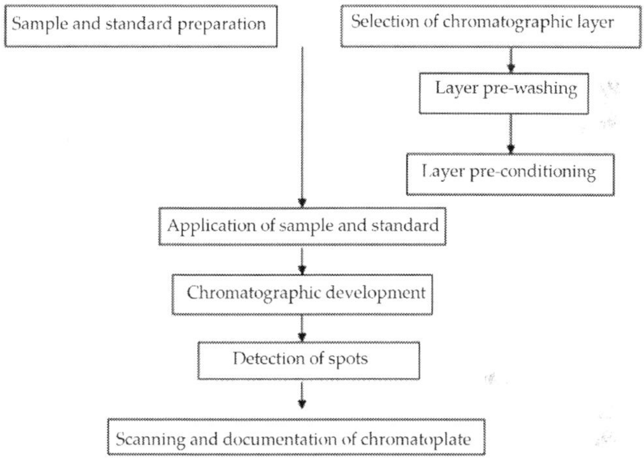

Figure 4.5: Schematics for HPTLC

SAMPLE PREPARATION AND APPLICATION

Without interfering with the solvent front, an ideal solvent aids in the resolution of samples. The peaks of interest should be resolved between Rf 0.15 and 0.85. The elution power of the mobile phase depends on a property called eluent strength which is related to the polarity of the mobile phase components. The more nonpolar the compound, the faster it will elute (or the less time it will remain on the stationary phase) and the more polar the compound the slower it will elute (or more time on the stationary phase).

Pharmaceutical preparation with sufficiently high concentration of analyte is simply dissolved in a suitable solvent that will completely solubilize the analyte and leave excipients undissolved to yield a test solution that can be directly applied on HPTLC plate. It is a fact that application of the sample is the most critical step to obtain good resolution for quantification in HPTLC21.Sample application technique depends on factors such as the type of sample matrix, workload and time constraints.

Although chromatogram development is the most crucial step in the HTLC procedure, important parameters are generally overlooked. 28 HPTLC plates are developed in twin-trough chambers, or horizontal-development chambers.

In general, saturated twin-trough chambers fitted with filter paper offer the best reproducibility. Twin through chamber avoids solvent vapor preloading and humidity. 40-42 Detection Detection of separated compounds on the sorbent layers is enhanced by quenching of fluorescence due to UV light (ranged normally at 200-400 nm). This process is commonly called Fluorescence quenching.

QUANTITATION:

Generally quantitative evaluation is performed by measuring the zones of samples and standards using a densitometer or scanner with a fixed sample light beam in the form of a rectangular slit. The chromatogram can be scanned in reflectance or in transmittance mode by absorbance or by fluorescent mode; scanning speed is used up to 100 mm/s. 23,28 Due to scanning spectra calibration of single and multiple levels of linearity, linear and nonlinear regression equations are possible. Scanning has been done by two methods i.e. Slit Scanning and Video Scanning.

Slit Scanning densitometry (SSD)

SSD is now relatively mature and although limited to absorption and fluorescence detection in the UV–visible range. It consists of fiber optic bundle for illumination of sample zones and collection of reflected light (or fluorescence). Photodiode-array detector is used for simultaneous length detection and spectral recording. Video Scanning.

Video densitometry (VD)

VD is fast and simultaneous data acquisition from the whole plate. In which optical scanning takes place electronically, using a computer with video digitizer, light source, monochromators, and appropriate optics to illuminate the plate and focus the image onto a charge-coupled device (CCD) video camera. It is also useful in two-dimensional separations for thin-layer chromatography with image analysis

Advantages of HPTLC

- A better and more precise analysis is reported with the running of samples with standards depending the use of an Internal standard.
- Several analysts work simultaneously
- Lower analysis time and less cost per analysis
- The maintenance cost is lower.
- The samples do not need tedious preparatory steps and samples of several classes can be applied.
- The solvents do not require pretreatments.
- The amount of solvent required for a sample is very low.
- As each analysis uses a new system there are no chances of contamination in the analysis.
- The use of an open detection system permits visual detection.
- The compounds that do not absorb UV can be detected by post-chromatographic derivatization

APPLICATIONS

The technique of HPTLC finds application in several activities such as establishing the identity of molecules or analysis of purity along with estimation of concentration of a sample. Applied with latest technology for detection, elution and solvents along with chemical modifications contribute to the utility of the HPTLC methodology. These bestow on it an edge over other techniques such as GC or HPLC (5).

Few advantages associated with the technique over TLC include:

- Reproducible results.
- Precise results
- Documentation of results is facilitated.

TLC in pharmaceutical analysis

Because of this, HPTLC technologies are also the most appropriate TLC technique for conformity with GMPs (6). Today the comprehensive use of ***TLC in pharmaceutical analysis*** is demonstrated by the great number of artiicles published in this field (7).

Isoflavones are derivatives of 3-phenylchromen-4-one (8). They are found in many plants and plant-derived foods in both as aglycon forms and as β-glucosides, acetyl- or malonylesters1. Red clover (*Trifoliumpratense* L.), kudzu (*Puerarialobata*, Willd.), soy (*Glycine max* L.) and soy products such as soymilk, tofu or miso are especially rich sources of these polyphenols. Recently, these group of natural compounds have gained widespread attention as potential therapeutic agents because of their antioxidant, anibiotic, anti-inflammatory and anti-allergic properties.

The most employed method for identify and quantify of isoflavones is high-performance liquid chromatography (HPLC)7-10, however its main disadvantages are long analysis times, sophisticated instrumentation and high amounts of solvents during the whole validation process. On the other hand, HPTLC is useful and inexpensive tool for the phytochemical assessment of plant extracts and herbal drug formulations. It is willingly used especially as a screening method due to high sample-throughput and simple samples preparation without time-consuming process of purification. The retention behavior of commonly occurring compounds with estrogenic activity: daidzein, formononetin, genistein, prunetin, biochanin A and coumestrol on silica and silica modified with cyjanopropyl, aminopropyl, diol and octadecyl groups was investigated.

Mixtures of methanol or acetonitrile with water and eluents such as chloroform, heptane and toluene with more polar modifiers: ethyl acetate, acetone, diisopropyl ether in various concentrations were used as mobile phases. The best result of separation was obtained on RP-18s plates with use diisopropyl ether/toluene/formic acid (40/60/0.5 v/v/v) as eluent. All investigated compounds were well separated which enabled their densitometric analysis. The proposed chromatographic system was used to qualitative analysis of investigated compounds in plant material.

Clinical Studies

Le Roux *et al (1992)* studied the use of HPTLC for clinical studies (9,10). The technique was applied to quantify the level of salbutamol in serum samples. The technique of HPTLC also has facilitates the separation of more than twenty families of lipids. The technique finds mention as a tool in study of clinical samples (11). The technique of HPTLC was applied in the analysis of pills that included nevirapine, lamivudine and

Stavudine (12). The determination of 2 protocols that described the use of HPTLC aided in the assays of key component of Mexicord drug. The method showed advantages in the following aspects:

- Results were linear
- The readings were accurate and precise.
- The detection was specific.

HPTLC to Study Natural Products

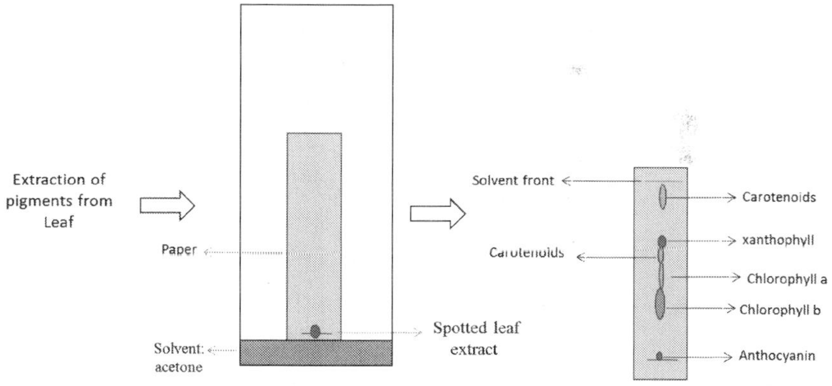

Figure 4.6: Separation of plant pigments by TLC

The technique of HPTLC shows versatility, rapidity and sturdiness as a technique of detection. Apart from checking for impurities, the technique can also screen the various stages of harvest or growth. For instance, a 2011 study used HPTLC to measure the amount of diterpenoids in the root bark of *Photiniaintegrifolia.*

Using markers that belonged to the family of diterpenoids, HPTLC was shown to be effective in the assay to yield results with high reproducibility (13).

Analysis of stem cell lipids

Apart from the key role of MALDI-TOF MS in protein studies, the analysis of lipids with the instrument has also been reported. Depending on the applied matrix, however, some lipid classes are more sensitively detected than other ones and this may even lead to suppression effects if complex mixtures are analyzed (14). This necessitates a prior classification of such lipid samples.

Lipids that were resolved on a HPTLC plate could be subjected to analysis using MALDI-TOF MS using extracts of tissues and mixtures of lipids.

Dongre and Kamble (2003) reported the application of HPTLC for detection and identification of heroin samples (15). The detection of the diacetylmorphine was performed by a mixture of dipyridyl and iron chloride based chemical. The reagent caused the development of spots of red color when heat was applied.

This detection could also be applied to other molecules such as codeine, morphine that has the potential to be extended into the realm of forensics.

REFERENCES

1. Gunter. Zweig and Joseph. Sherma. Paper and thin-layer chromatography. *Anal. Chem.*, 1980, *52* (5), pp 276–289
2. Theory and Mechanism of Thin-Layer Chromatography. n book: Handbook of Thin-Layer Chromatography, Edition: 3rd Edition, Revised and Expanded, Chapter: Chapter 2, Publisher: Marcel Dekker, Editors: Joseph Sherma and Bernard Fried, pp.47-80
3. Vidhate GJ, Todkar VD, Tare HL, Bidkar SJ, Bidkar JS, Dama GY. HPTLC A Technology of Industrial Utility. Int J Institutional Pharm Life Sci. April 2017; 7(2).

4. Shivatare RS, Nagore DH, Nipanikar SU. 'HPTLC' an important tool in standardization of herbal medical product: A review. J Sci Innov Res. 2013; 2 (6): 1086-1096

5. Attimarad M, Ahmed KKM, Aldhubaib BE, Harsha S. High-performance thin layer chromatography: A powerful analytical technique in pharmaceutical drug discovery. Pharm Methods. 2011;2(2):71-75. doi:10.4103/2229-4708.84436.

6. Sudberg S, Sudberg EM, Terrazas J, Sudberg S, Patel K, Pineda J, Fine B. Fingerprint analysis and the application of HPTLC to the determination of identity and quality of botanicals, from an industry perspective. J AOAC Int. 2010 Sep-Oct; 93(5):1367-75.

7. Arup U, Ekman S, Lindblom L, Mattsson JE. High performance thin layer chromatography (HPTLC), an improved technique for screening lichen substances. Lichenologist. 1993;25:61–71.

8. Wójciak-Kosior M, Sowa I, Strzemski M, Rokicka K, Blicharski T, Bogucka-Kocka A. Application of High Performance Thin Layer Chromatography For Qualitative Analysis Of Isoflavones In Medicinal Plants. International Journal of Research In Pharmacy And Chemistry. 2014, 4(1), 222-227

9. Le Roux AM, Wium CA, Joubert JR, Van Jaarsveld PP. Evaluation of a high-performance thin-layer chromatographic technique for the determination of salbutamol serum levels in clinical trials.

10. J Chromatogr. 1992 Oct 23; 581(2):306-9.

11. Bernardi T, Tamburini E. An HPTLC-AMD method for understanding the metabolic behavior of microorganisms in the presence of mixed carbon sources. The case of Bifidobacterium adolescentis MB 239. JPC-J Planar Chromatogr-Modern TLC. 2009;22:321–5

12. Shewiyo DH, Kaale E, Ugullum C, Sigonda MN, Risha PG, Dejaegher B, Smeyers-Verbeke J, Vander Heyden Y. Development and validation of a normal-phase HPTLC method for the simultaneous analysis of lamivudine, stavudine and nevirapine in fixed-dose combination tablets. J Pharm Biomed Anal. 2011 Feb 20; 54(3):445-50.

13. Yadav D, Tiwari N, Gupta MM. Simultaneous quantification of diterpenoids in Premnaintegrifolia using a validated HPTLC method. J Sep Sci. 2011 Feb; 34(3):286-91.

14. Fuchs B, Schiller J, Süss R, Zscharnack M, Bader A, Müller P, et al. Analysis of stem cell lipids by offline HPTLC-MALDI-TOF MS. Anal BioanalChem. 2008;392:849-60. 36.
15. Dongre VG, Kamble VW. HPTLC detection and identification of heroin (diacetylmorphine) in forensic samples. Part III. J Planar Chromatogr-Mod TLC 2003;16:458-60

CHAPTER 5

GAS CHROMATOGRAPHY

INTRODUCTION

Gas chromatography (GC) is one of the advanced systems of chromatography. It is a technique for the separation of components of a mixture especially for volatile molecules. It was invented by Martin and Synge in the year 1941 (1) and was tested for the separation of volatile fatty acids (2).

As of date, GC is reported as a vital and universal approach in the field of analytical chemistry (3).

It employs a method of variations in sequestering compounds on the basis of a stationary phase and a mobile phase. The technique has been reported to be sensitive, effective and simple (4).

A central theme of chromatographic experiments is the column into which samples are applied. The field of research in developing new materials for columns is constantly coming up with new innovations.

PRINCIPLE

The basic tenet of partition chromatography finds expression in the rationale of gas chromatography. A difference in the separation or partition of sample molecules between the stationary phase in the column and a mobile phase leads to resolution of sample molecules. The technique of GC involves the application of a sample on a column used for separation. The mobile phase in this technique is an inert gas: based on the associations between the molecules in the sample and the stationary phase particles, molecules are separated. A subsequent step of elution, the molecules that have separated are detected in the form of a signal that is in proportion with its concentration (4).

PREPARATION OF SAMPLES

In order to enrich the amount of volatile materials, GC involves preparation of samples to circumvent the contamination by molecules of high molecular weight.

Distillation techniques use differences in the properties of samples such as vapor pressure. Depending on the samples, steam distillation or water-steam distillation or hydrodistillation may be performed followed by addition of a dying agent such as sodium sulfate. Instead of traditional heating systems, the development of Microwaveassisted hydrodistillation (MWHD) for solid samples has made the process more rapid.

Extraction techniques use the principle of preferential interaction onto a solid surface (porous such as resins or Porapak or microporous such as molecular sieves or alumina) or dissolution of an analyte in a solvent. Liquid-liquid extraction for liquid samples were extensively used while Solid-phase extraction serves as a replacement for liquid extraction as it circumvents the need of removal of solvent. Supercritical fluid extraction eliminates the presence of the agent used for extraction. Carbon di oxide finds a major application in this technique as the gas returns to its original form under optimum conditions leaving the sample.

Head space methods refer to extraction of samples from the vapor phase above the sample. This technique overcomes the presence of additional data generated due to solvents. There are 2 approaches in this method that include:

Static headspace is a technique where the headspace volume is applied on the GC column. This finds application in the analysis of volatile chemicals in water or paint or pharma products. However, dynamic headspace is applied when the analyte is present in low concentrations. The technique involves the flushing of the headspace with a non-reactive gas following which the analytes are transferred to a suitable medium like a solvent or adsorbent to be recovered by heat or solvents (5).

INSTRUMENTATION

The system consists of a column on to which a support or stationary phase is applied: it can be either a solid or a liquid. The mobile phase is a gas that flows from a source or cylinder through the column. When the sample is applied or injected into the column, the gas serves as a transport vehicle for the sample across the column. Based on the partition of the sample molecules between the flowing mobile phase and stationary phase packed in the column, resolution is achieved. A molecule that partitions on the support is eluted later while a molecule that partitions into the gas phase is eluted in the initial steps. A detector is responsible for the identification of the eluted molecules that is subsequently interpreted in the form of a chromatogram.

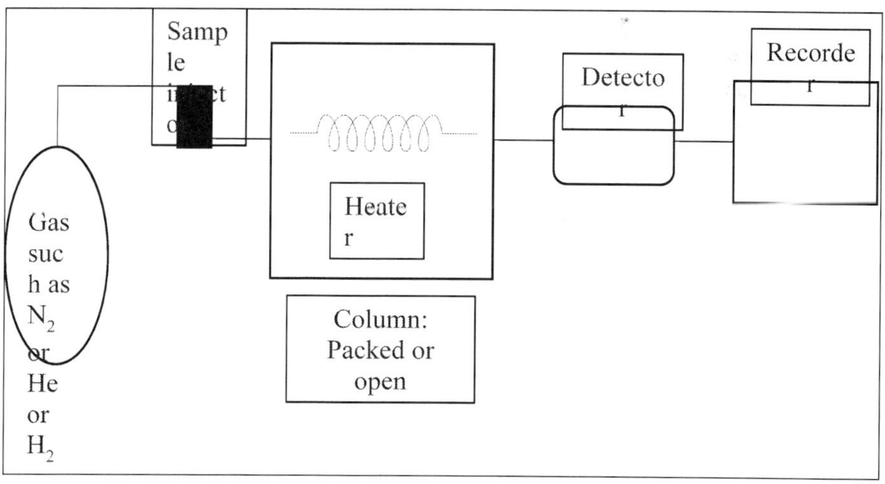

Figure 5.1: Schematic of Gas Chromatography

Each of the components can be summarized as follows:

The use of an ***autosampler*** introduces technology in the instrumentation to circumvent the use of manual application of samples. This equipment has been associated with an improvement in the time run along with reproducible results (6).

MOBILE PHASE

As previously mentioned, the mobile phase is a gas: an inert gas that is stored in a cylinder. The gases include Helium, Nitrogen and Hydrogen. Based on the flow rate for the experiment, the gas is allowed to flow from its container into the chromatography column. Each of the gases have a set of merits and demerits such as: Nitrogen gas is cheap and easily available while its flip side is an increase in the duration of the run time. Hydrogen is affordable and decreases the run time but is explosive while Helium shows a good effect with safety, the only challenge being its cost.

The instrument is equipped with molecular sieves to eliminate the presence of water. The most common gas is Helium as it is not combustible and can be used with ease. However, a boost in the cost has propelled users to examine Hydrogen as an alternative (6).

Table 5.1: Summary of the various gases employed in the chromatography process.

Gas	Pros	Cons
Nitrogen	It is inexpensive and available with ease.	It has been reported to increase the run time.
Hydrogen	The gas is very inexpensive and offers lesser run times.	The gas is explosive.
Helium	The gas is safe to use and is a good compromise.	The gas is expensive.

INJECTORS

These components are used for application of samples into the column to achieve their resolution. An example of an injector is a Septum type injector.

A glass tube is the major part of the injector, into which the sample is applied via a rubber septum that can seal spontaneously and is made of silicone. The sample is subjected to vaporization inside the injector: hence, an important point to be considered is that the conversion of the samples to gas should occur in minimum time that necessitates the use of elevated temperatures. This ensures a good resolution while the lack of high temperature can cause the bands to become aberrant.

As the sample traverses across the septum, it undergoes vaporization on account of the elevated temperature to be transported by the mobile carrier gas.

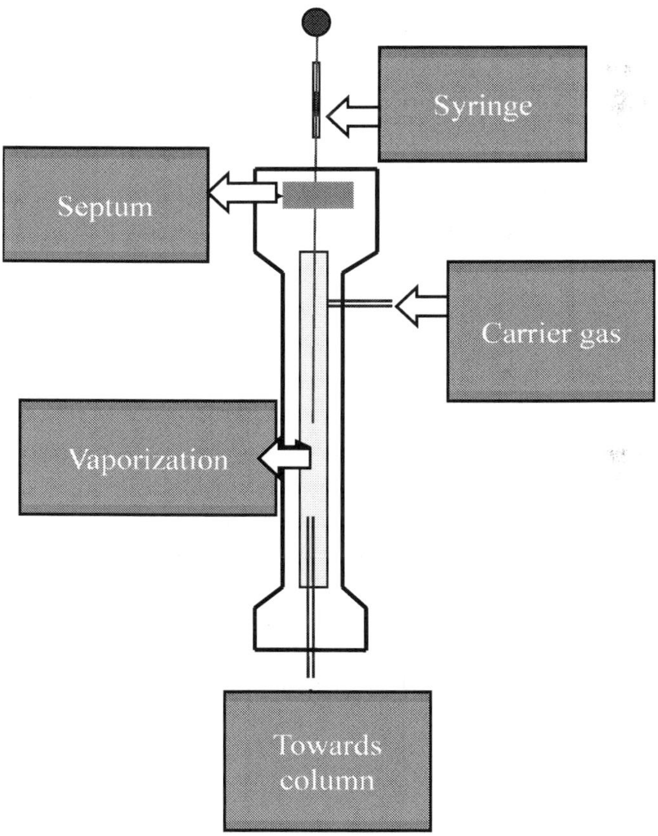

Figure 5.2: A septum injector

The inlet is attached with the column head. The common inlets are: S/SL (split/splitless) injector, on- column inlet, PTV injector, and Gas source inlet or gas switching valve, P/T (Purge-and-Trap) system (6). These systems can be operated in 2 mechanisms:

***Split*:** where samples of low dilution are applied in small quantities.

***Split less*:** the entire amount of sample is applied to the system and is used for samples that have low amounts of analytes.

Care should be exercised in the injection of sample in the form of a thin band to prevent aberrant peaks. An added condition is the various molecules in a sample must be prevented from reaction with each other or subject to chemical transformations. The temperature at the inlet must be controlled as very low temperatures may hamper the volatilization while elevated temperatures may cause destruction of samples (5).

COLUMNS

The columns support the separation of components of a mixture and are placed within the thermal chamber or oven that requires their coiled conformation.

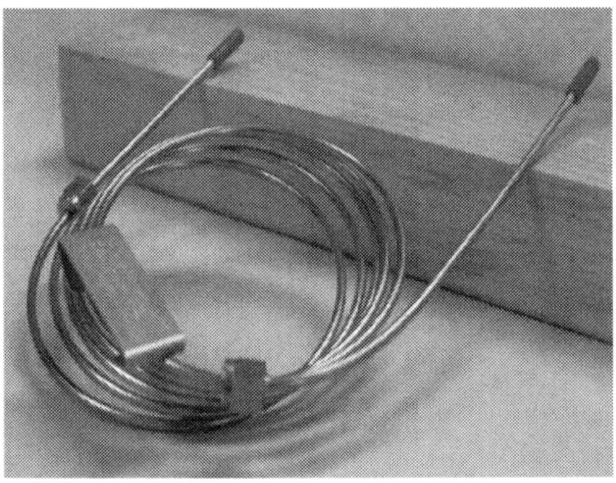

Figure 5.3: Coiled column

The columns are of several types:

Packed GC columns

These columns are generally stainless steel with reports of other materials that include: nickel, glass, Teflon or a combination of stainless steel and Teflon. These columns were employed in the initial years in the instruments that were designed previously.

Dimensions of these columns are reported as 0.61–3.05 m length and 0.32 to 0.64cm in their diameter. These columns are packed with support that contains the stationary phase: a proportion of 5 to 20% of the phase has been reported. The solid support used should have certain requirements such as:

- It should be inert or non-reactive.
- Its surface area must be large: 0.5–5 m^2/g.
- It should be mechanically strong to withstand the system during the sample application or operation of the system.
- It should also display minimum adsorption towards the analytes.

An example of an inert support is Chromosorb® (7) which is a combination of Diatomaceous earth along with silica in a hydrous state.

Diatomaceous earth possesses a large number of active sites by virtue of the presence of -OH groups that react with many molecules to cause aberrant peaks. This limitation necessitates the washing of the support with chemicals to deactivate them: Gas Chrom Q II®, Chromosorb W-HP®, Supelcoport® are examples of such supports.

Alternatively, liquid phases can be used that include methyl trifluoropropyl silicone, methyl silicone, methyl cyanoethyl silicone or methyl phenyl silicone. These oils are derived from silica and are stable over high temperatures. Liquid phases find application especially at elevated amounts to minimize interactions between molecules with solid supports (4).

Capillary GC Columns

These columns consist of a coating of liquid phase in the form of a film on the inner surface of a tube. The tube is not sealed and offers minimum resistance to the flow of the mobile phase. Such systems are called open tubular columns on account of these properties.

The columns are of the following types ***based on the construction***:

Wall-Coated Open Tubular Column (WCOT)

The columns contain a coating of the stationary phase as a thin layer on the surface of the wall of the tube. The coating thickness is reported at 0.05–3 μm (4).

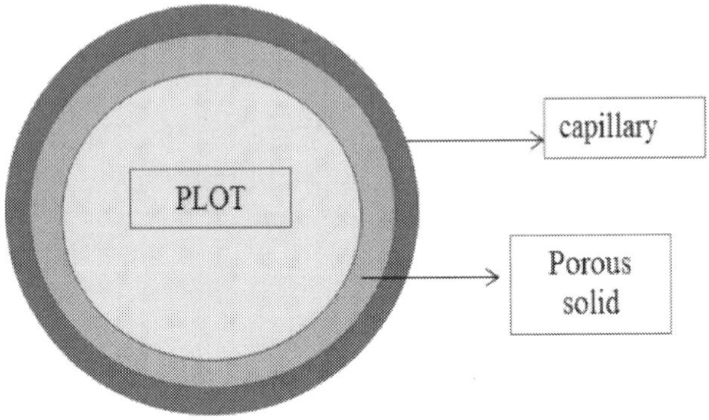

Figure 5.4: WCOT

These columns were initially described in 1950s as hundred times more efficient than packed columns (8).

Porous Layer Open Tubular Column (PLOT)

These columns possess a layer of porous materials that is capable of adsorption. Materials used commonly include molecular sieves or alumina or Porapak.

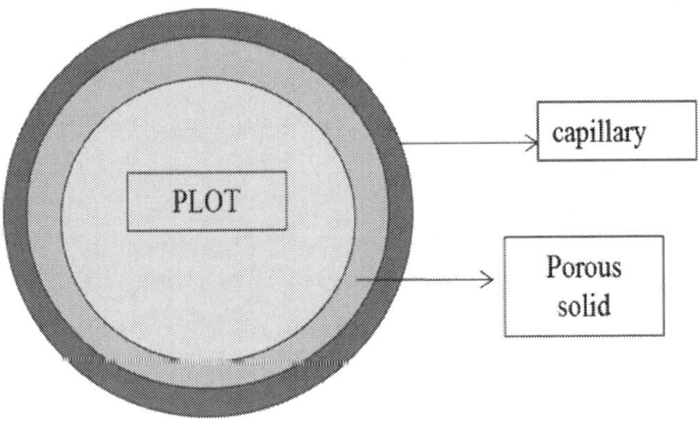

Figure 5.5: PCOT

These columns are recommended for the study of volatile materials inclusive of gases and have been much used and modified for use (4).

Support-Coated Open Tubular Column (SCOT)

These tubes have the liquid phase that is applied on a solid support that is adsorbed on to the capillary. Supports used can include materials like celite. These columns can accommodate a larger quantity of liquid phases than compared to earlier models of WCOT that possessed thin wall coatings of the phase. The development of WCOT with thicker coatings of liquid phase has put this column in the backseat (4).

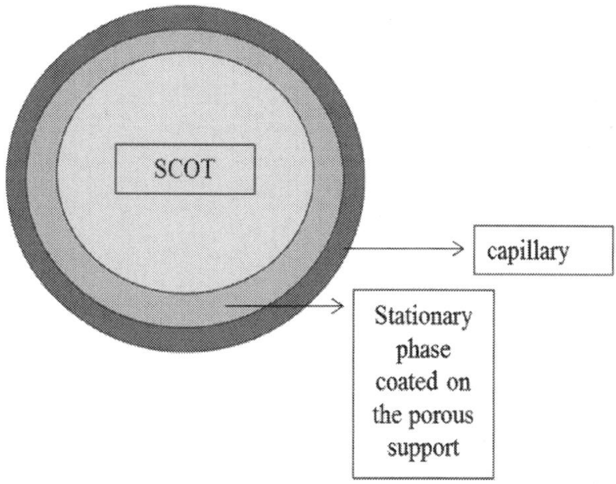

Figure 5.6: SCOT

Fused Silica Open Tubular Column

The capillaries used in GC experiments were initially composed of stainless steel that had certain limitations: such as its unwanted interactions with biomolecules or the reports of the inefficient output. This led to the use of fused silica as capillary columns that find wide use today (9,10).

The use of these columns based on fused silica are preferred due to smaller dimensions than glass columns with advantages of flexibility to be coiled and are inert to a large extent (4).

The capillary based GC columns are used to a large extent in the industry with reports of 80% of applications using the technique. The technique is more advantageous than compared to packed GC columns especially in terms of efficiency. The non-uniform and thick, packed columns yield an output of 2000 plates per meter while capillary columns especially the ones based on fused silica offer 3000–5000 theoretical plates per meter.

THE PERFORMANCE OF A COLUMN IN GC

This is defined by the numbers of theoretical plates:

$$N_{eff} = \left(\frac{t'_R}{\sigma}\right)^2$$

Where N_{eff}: is numbers of theoretical plates

t'_R is the adjusted retention time

σ is the standard deviation of the peak

Another fall back of packed columns is there packing with solid support that prevents the use of long columns. On the contrary, capillary columns offer little resistance to flow due to their open nature to allow long columns for operations (4).

Table 5.2: A summary of the differences in the columns.

Capillary column	**Packed column**
The stationary phase is applied on the inner surface of the tube.	The stationary phase is applied on the column.
These tubes are not sealed i.e. open and hence offer little resistance to flow.	As the tubes are sealed, they present pressure drops.
They allow long operations	They cannot permit long operations.
They are reportedly more efficient due to increased lengths.	As their length is limited, efficiency is less.

Gas - Solid Chromatography (GSC)

The technique works on the principle of adsorption of a gas on a solid. It employs the use of a stationary phase such as alumina or silica along with a carrier gas as the mobile phase. As the distribution coefficients in this method are higher than when liquids are used, it can be used for the resolution of samples that cannot be applied on liquid systems.

This includes the resolution of air components or oxides of Nitrogen or Hydrogen Sulfide or oxides of carbon and methane.

Gas - Liquid Chromatography (GLC)

The technique is applied for the separation of volatile substances and uses a stationary phase that is a non-volatile or has low volatility while a gas serves as the mobile phase.

In general, the dimensions of the stationary phase play a key role in the output of a GC run. The increase of thickness of a system has a key role in:

More numbers of samples can be resolved

The increase in dimensions causes a boost in the retention time which in turn causes efficient resolution of volatile analytes.

The separation of small and large molecules can be accomplished with the use of molecular sieves. These are exchangers composed of metal aluminum silicate that have several pore sizes according to the analytes to be separated.

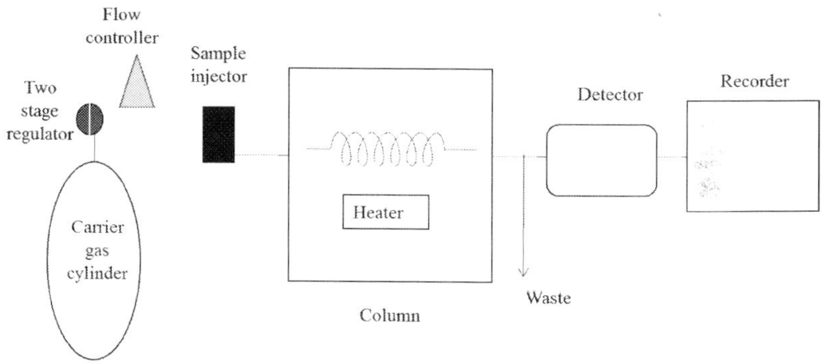

Figure 5.7: The set-up of GLC

THERMAL CHAMBER

This component of the GC apparatus is responsible for the application and maintenance of temperature. As the columns and detectors are sensitive to fluctuations in temperature, the thermal unit is responsible for an optimum temperature. An elevated temperature is responsible for the volatility of samples to ensure smooth flow and ultimate resolution.

DETECTORS

Detectors can include mass flow based ones that does not depend on the gas used as carrier while concentration dependant detector observe the solute in the sample without affecting the sample and is affected by the gas used (6).

Several types of detectors are employed in GC that includes:

- Thermal conductivity detector (TCD)
- Flame ionization (FID)
- Atomic-emmision detector (AED)
- Chemiluminescence detector
- Electron-capture detector (ECD)
- Flame-photometric detector (FPD)
- Mass spectrometer (MS)
- Photoionization detector (PID)

Thermal conductivity detector:

It comprises a wire that is heated by current connected to two columns. A sensor measures the conductivity of the gas in the chamber. The instrument measures two gas streams, one stream which is only the carrier gas, the other one comprising the transporter gas and the compound.

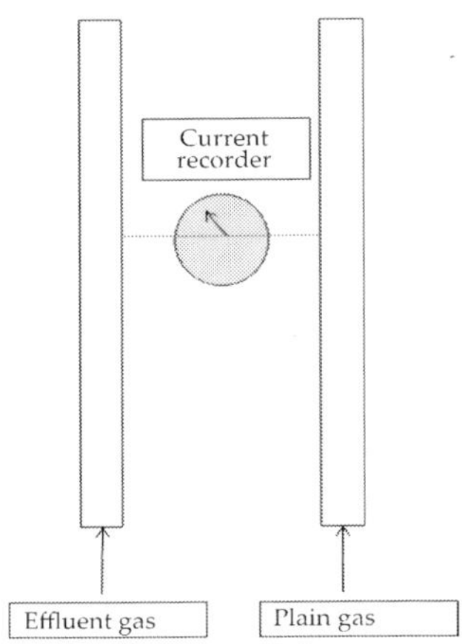

Figure 5.8: Thermal detector

When the gas flows through the columns, due to constant temperature, the detection does not occur while the application of sample causes a shift in temperature that is recorded. Depending on the sample, a shift is recorded.

A Wheatstone bridge circuit records the temperature distinction between the reference and the samples. The sensitivity of the instrument is 10^{-5}-10^{-6} g/s.

Flame ionization detectors:

They are reported to be highly sensitive especially for organic moieties (10-12 g/s = 1 pg/s) but less tolerant to small inorganic groups. Using a flame in the chamber, samples are ionized from the effluent. These ions migrate towards appropriate electrodes based on charge. The interaction of an ion with an electrode generates a signal proportional and specific to the sample in question.

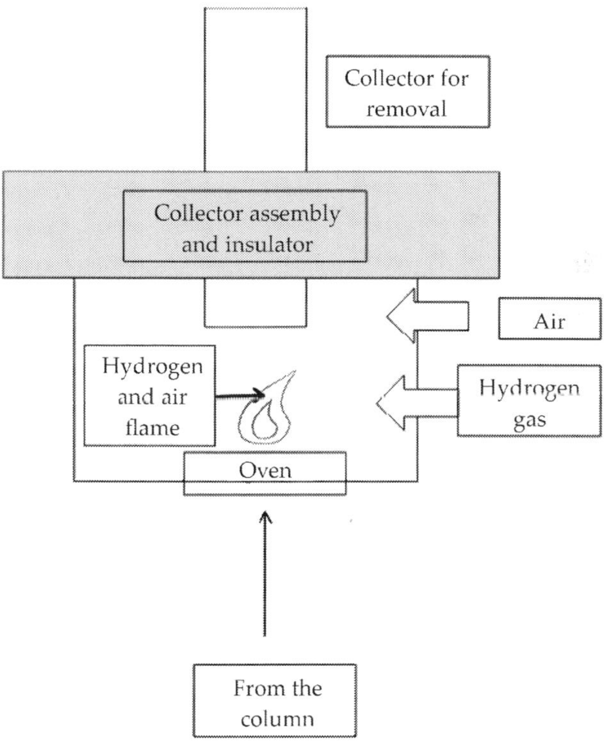

Figure 5.8: Flame ionization detector

A few points to be noted include the lack of quantification between the signal and carbon numbers apart from the determination of reaction molecules for an instrument that is used. The technique also entails the use of several gases in the chamber.

Argon ionization detector:

Using a similar principle of flame ionization detectors, these units employ argon instead of hydrogen in the ionization of samples.

Electron capture detector or ECD Detector:

The unit employs a source of radiation: radiactive Beta emitter auch as 63Ni, 3H that causes ionization of samples present in a carrier gas. The interaction between the beta particles and carrier gas and samples produces a burst of ions. The presence of electornegative functional groups causes an interaction between the electrons and the groups to cause a dip in the current signal that are recorded on a Wheatstone bridge.

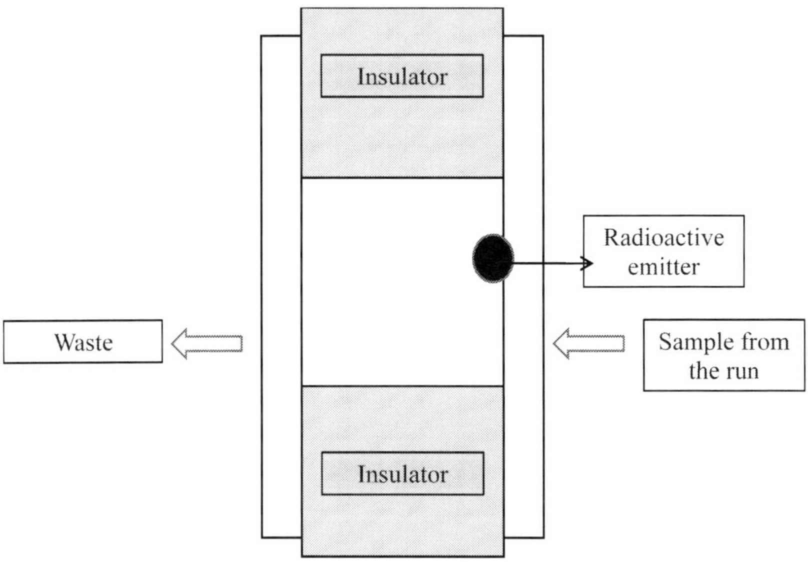

Figure 5.9: The ECD

This technique finds application in evaluation of electronegative group containing compounds such as polychlorinated biphenyls or pesticides.

MS Detector:

A combination of GC and MS offers advantages of both techniques with GC separating the analytes, and the mass spectrometer distinguishing patterns of fragmentation. The technique employs the difference in mass-to-charge ratio (m/e) of ionized species to enable the separation. Based on structures of the species generated, different patterns of fragmentation are observed for the samples. The technique involves: the creation of ions in gas phase followed by their separation using mass to charge ratio. The separation of ions in the experiment is quantified as: R = m/Dm where: m is the ion mass and Dm is the difference in mass between two resolvable peaks in a mass spectrum.

Photo-ionization detector (PID):

The system records the progressions of a sample to methane and studies the pattern of the samples with equivalence to that of methane. The technique can be applied to complex mixtures of carbon frameworks with the lack of standards.

Nitrogen-phosphorus detector:

It works on the principle of the change in voltage of a heated bead upon sample elution that is induced by the presence of nitrogen and phosphorus in the system.

Alkali Flame Detector or AFD:

The system uses alkaline metal particles that are flooded with Hydrogen, in the absence of heat, the system is similar to FID.

Flame photometric Detector (FPD):

The technique uses a photo multiplier to analyze the spectrum of sample analytes as they are heated. The sample molecules from elution are heated with Hydrogen to cause excitation of several groups on a sample. Each excited moiety emits light that is dependent on the molecule (for instance: sulfur emits at 394 nm) and is measured by a photomultiplier (6, 11).

Table 5.3: The samples and detection of various carrier systems are discussed in the following table:

Carrier system	Detection limit	Samples
Thermal conductivity	500pg in 1ml	It can be applied to a range of samples.
Flame ionization	1pg in a second	Hydrocarbons
Electron capture	5pg/s	It can be applied for compounds that contain halogens
Mass spectrometers	0.25 to 100pg	The technique can be applied to many samples.
Photo ionization	2pg of Carbon per second	It can be applied to compounds that can react with light
Fourier Transform	0.2 to 40mg	It is exclusively for organic compounds.

THE RETENTION INDEX

The identification of solutes in a chromatogram was described by Kovats in the 1950s. The parameter can be applied to a sample using a retention time of two alkanes that have similar retention times as the sample. The alkanes in question should have more than 4 carbon atoms and are used as standards to build the scale (figure 5.10). The retention index is defined as "100 times the quantity of carbon atoms of an analyte is the index for an alkane". It is not dependent on the conditions that prevail during an experiment. For a particular stationary phase, the retention index of a compound is constant but varies with other stationary phases.

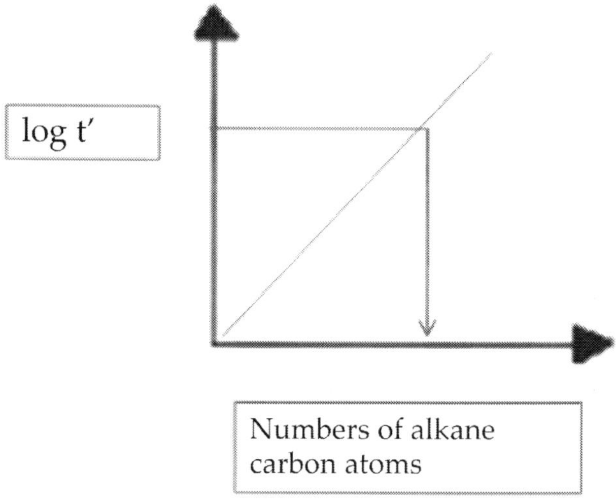

Figure 5.10: The retention index of an alkane

The adjusted retention time, t_R’, is defined as:

$t_R' = t_R - t_M$

where tR is retention time of an analyte

tM is the time of elution for a peak that is not retained

With the plot of logarithm of the adjusted retention time against the number of carbons of standard alkanes, the index can be calculated (12, 13)

DERIVATIZATION AND GC

This entails the conversion of a sample analyte through reactions for subsequent detection on GC. There are several purposes of derivatization that include: preventing the decomposition of polar molecules due to heat to enable elution at optimum temperatures. Another challenge is the presence of a reactive Hydrogen on certain analytes (in the form of -OH, -SH,-COOH etc.) .Such hydrogen atoms participate in the formation of intermolecular hydrogen bonds that affect the performance in the column.

The pre-preparation before GC can alter the state of volatility of a compound: either elevate or depress the volatility. Several techniques of derivatization exist that include (14):

Alkylation

This is usually an initial step for other derivatization reactions and involves the substitution of an aliphatic group in place of a Hydrogen. Many reagents are employed for this purpose, a few including: Dialkylacetals, Diazoalkales, Pentafluorobenzyl bromide (PFBBr), Benzylbromide, Boron trifluoride (BF3) in methanol or butanol and Tetrabutylammonium hydroxide (TBH).

Acylation

This reaction is performed to confer protection to sensitive groups and increase stability of an analyte and also facilitates detection on Electron capture detectors as they introduce such groups on samples. Reagents used include: Fluoracylimidazoles, Fluorinated Anhydrides, N-Methyl-bis(trifluoroacetamide) (MBTFA), Pentafluorobenzoyl Chloride (PFBCI) and Pentafluoropropanol (PFPOH).

Silylation

This modification is performed to render a non-volatile molecule volatile to be loaded on a GC system. Reagents used for the silylation derivatization process include Hexamethyldisilzane (HMDS), Trimethylchlorosilane (TMCS), Trimethylsilylimidazole (TMSI), Bistrimethylsilylacetamide (BSA), Bistrimethylsilyltrifluoroacetamide (BSTFA), N-methyltrimethylsilyltrifluoroacetamide (MSTFA), Trimethylsilyldiethylamine (TMS-DEA), N-methyl-N-t-butyldimethylsilyltrifluoroacetamide (MTBSTFA), and Halo-methylsilyl derivatization reagents.

LATEST RESEARCH IN MATERIALS USED IN GC COLUMNS

The conventional materials that were used as columns included polysiloxane after which polyethylene glycol made an appearance as a stationary phase. The use of such a stationary phase possesses challenges such as the limit of temperature they can withstand, another limitation in the extent of their modification to expand the selectivity of the column and unwanted reactions in the presence of air or moisture (4).

A new material termed as Organic Ionic Liquids (IL) were tested in the 1980s. Using ethyl ammonium nitrate, the team could achieve the resolution of a several compounds at low temperatures: below 393K (16). As the compound was found to deliver less efficient results, a subsequent compound reported included Ethylpyridinium bromide. It was reported that the compound was as efficient as the materials used traditionally as column packing (16).

- The use of these Ionic Liquids presents several advantages that include:
- Their properties: physical and chemical such as viscosity, can be tailor made or designed: a unique feature that makes these compounds stand apart. The addition of chains or chemical alterations can exert a control on the experimental set up and hence the outcome of the separation.
- The compounds can be subjected to immobilization as well as cross linking.
- The compounds can be used in the resolution of both non-polar and polar compounds.
- The compounds developed are stable over high temperatures (4).

This was followed by the development of ***metal-organic frameworks*** that were reported as porous substances: inorganic structural units with organic linkers along with moieties that can chelate metals (17). These materials can be applied in the form of sieves especially in a tandem format to be used for resolution of complex materials and also micro-extraction. The materials possess certain advantages that favour their use in experiments that warrant high resolving: they are easily synthesized and are stable over high temperature (18).

A study in 2005 reported that a chromium terephthalate metal-organic framework was found to have easily available co-ordination centres apart from big pores and an enhanced area apart from being stable to several chemicals and ranges of temperature (19).

Another novel material to enter use as stationary phase is ***carbon nanotubes*** that have been reported to possess desirable properties in the physiochemical and electrical arena (20). These materials possess a property of π stacking in conjunction with several organic molecules apart from the presence of abundant active sites (4).

A material, ***graphene oxide*** is synthesized as a single layer of nano proportions, from graphite and is reported to possess properties such as:

- Thermal stability
- Big area
- Good adsorption power
- Dissolves in aqueous (4).

MICRO GC

The development of microelectromechanical systems (MEMS) technology has seen a surge with a view to circumvent the use of heavy aspects of instrumentation. The major huge material in GC is the column itself. With a view to develop micro columns, a 2005 work published silicon glass microgas chromatography columns. This system was found to have high performance and included sensors and heaters. The apparatus was also found to fulfil the conditions of pressure and temperature along with a reduction in the requited power. As compared to systems with constant energy, the susyem separated 20 molecules in a mixture in half the time required (21).

APPLICATIONS OF GC

The technique of GC finds application in checking the purity of several organic samples as the presence of extraneous matter is detected as extra peaks. With the adherence to good practices, it is possible to achieve high reproducibility with the technique as the retention times can be calibrated.

The technique can also be used to measure or estimate the concentration of a sample under study. The concentration of an analyte is proportional to the area of its peak. Using this principle, a calibaration curve can be plotted by running the system using standards and subsequently used for the calculation of an unknown/novel analyte.

CASE STUDIES

Study of Indoles of Plants

Using the advantages of GC reported namely, the increased resolution as well as its sensitive detection, a team examined GC for analysis of alkaloids. The ability of the technique to be linked with MS (22, 23) aids in the analysis of novel metabolites in a sample. The use of GC in the characterization of several metabolites such as alkaloids has been reported. A group examined capillary GC for the evaluation of alkaloids of Aspidosperma namely: *A. pyricollum, A. olivaceum, A. pyrifolium, A. polyneuron and A. ramiflorum*. Using a small column of length 10cm, a wide range of alkaloids were separated using a capillary in a fast manner and high resolving power. The technique enabled the separation of the entire set of compounds of *A. olivaceum,* when the MS data was analysed it was seen that the patterns of the fragments and molecular weights were within the expected and reported values. The same technique yielded alkaloids such as 10-methoxy-geissoschizol, (E)-isositsirikine, ramiflorineA, e ramiflorine B and β-yohimbine from *A. ramiflorum.*

The experimental setup for GC involved homogenization of extracts and followed by filtration in a silica gel column followed by elution. The samples were applied on GC: Hewlett-Packard 3396A gas chromatograph equipped with a FID. The column details were: (12 m, 0.25 mm i.d. x 0.2 mm film thickness, J & W Scientific, CA) while th carrier gas was Hydrogen. The analysis was performed on HP 5970B chromatograph linked with an MS detection unit (23,24)

Identification and Determination of Metabolites in Plant Cell

The non-polar metabolites in a plant tissue, in this case: *Chrysanthemum cinerariaefolium* (pyrethrum) or *Tagetes* species (marigold) were studied using Gas-liquid chromatography (GLC) and gas-liquid chromatography/mass spectrometry (GC/MS). The techniques were applied for ascertaining the identity apart from examining non-polar metabolites in the tissue. The author used electron capture detector as it allowed the choosing of such plant tissue materials with an enhanced yield for subsequent scale up and commercialization. The experiment revealed six pyrethrins when leaf samples weighing up to 20mg were used for analysis. The electron capture detector allowed the quantification of α-Terthienyl in nano scale: with the smallest amount that was detected reported at 0.1 ng while the caliberation curves were linear between a range of 0.2-20ng.

Similarly, *T. erecta* root samples reported a level of 140 μg g^{-1} of α-terthienyl. The analysis of *T. minuta* leaf samples yielded a profile that covered 13 nonpolar compounds, of which 9 were identified. The molecules included 2 sesquiterpenoids, 3 thiophenes, 2 sterols with a precursor and ethyl linolenate. An additional data that was revealed was lack of synthesis of sesquiterpenoids or thiophenes in calli and cell suspensions while sterols were produced (25).

Determination of Biomarkers in Petroleum by Multidimensional Gas Chromatography

Crude oils when subjected to traditional GC-MS due to the presence of a large number of analytes that undergo elution simultaneously, display a hump pattern on a chromatogram. Such complexes that show this pattern are referred to as unresolved complex mixture (UCM) (26). As majority of the compounds in such samples are isomers, the co-eluting analytes are seen that poses a challenge in the analysis of such mixtures. This impediment is essential to overcome as such data offers a window to understand the evolution of the earth or oils studies. The chemical analysis and elucidation of the composition of crude oil is essential in the field of economics as the potential of use of heavy oils for use as appropriate commodities.

The use of techniques based on GC can be summarized as follows:

A 2001 study examined extra heavy gas oil by GC×GC-TOFMS to characterize the chemical composition of crude oil developed in marine environments. The authors identified several groups of molecules such as paraffins, hopanes, tricyclic terpanes, moretanes and steranes. Several other groups such as steranes, sulphur-containing aromatics, isoprenoids, alkylated aromatics, triterpanes were all reported using mass spectra data in combination with GC×GC. This data proved valuable to calculate several geological factors that include the migration of oils, biodegradation or thermal maturation patterns (27)

In other studies:

2D GC was applied on Ceará Basin Oils from Brazil to report the presence of essential oil components. The authors reported the successful separation with high resolving power to yield tricyclic terpanes/pentacyclic terpanes, C30/C30R demethylated homohopane, and hopanes/sterenes in the crude oil samples. The study was a first in the report of the identification of demethylated tri- and tetracyclic in Brazilian oils. This led to an important result that the compounds were set in due to some biodegradation and were not formed when sediment was transformed to sedimentary rock (28). examined oils from marine environments, GC×GC-TOFMS reported the presence of UCMs that were named as UCM, UCM I, UCM II, and UCM III. It was reported that predominantly mono- to hexacycloalkanes constituted UCM type I while C_{35}-C_{40} hydrocarbons constituted the UCM type II while UCMIII showed properties of both UCM I and II (28) in 2014 a study examined several biomarkers in petroleum to look at the amount of matter from terrestrials. The chemical characterization of several samples was done using GC×GC-TOFMS on oils from different basins in Brazil. The study unearthed several markers that were of plant origin that included lupanoides, olenoides, 18α-oleonane, and 18β-oleonane. Additionally, steranes, onocerane, hopanes, 8, 14-seco-hopanes, 3β- and 2α-methylhopanes were all isolated and determined. The oil from marine sources showed more amounts of 2α-methylhopanes while the 3β-methylhopanes and onoceranes were seen more in depositional lacustrine environments (29).

GC x GC was able to identify and separate several C_1-C_7 alkyl decahydrohopanes from oils isolated from several environments that included: land or sea sources (30).

GC And Oncology

The detection of cancer, while localized and early treatment regimes carried out, normally ends in higher clinical outcomes. GC can play a major position inside the profiling of the metabolism. A few examples on the role of GC and detection of biomarkers of cancer are highlighted:

Lung cancer:

A study examined levels of aldehydes in breath of sufferers with lung cancer via on-fiber derivatization strong-segment microextraction

(SPME)-GC-MS fifty-nine. Lung cancer is the main causative agent of most cancers loss of life within the international, and the survival rate is much less than 20% in a span of five years. On-fiber SPME offers many advantages over 'conventional' breath sampling strategies, which include pattern extraction and concentration in a single step and the said improvement of aldehyde stability. The technique was performed on forty patients with non-small cell lung cancer (NSCLC) and a 38 cohort 'normal' manipulate organization who showed no signs and symptoms, with 90% of NSCLC and 92.1% of controls being effectively diagnosed.

Colon cancer:

Metabolite profiling of human colon carcinoma through using GC-ToFMS changed into suggested in another study with detection of 206 metabolites with the aid of appearing a liquid-liquid extraction system sixty-eight. Of this wide variety, 107 may be recognized, with eighty four being registered inside the Kyoto encyclopedia of genes and genomes (KEGG) database and 71 being fundamental response partners in at the least one of the reactions annotated in KEGG reaction. The recognized metabolites have been believed to be associated with abnormalities in biochemical pathways, according to a brand new approach that calculates the space of every pair of metabolites inside the KEGG database interplay lattice. Paired samples of regular colon tissue and colorectal most cancers tissue have been differentiated at a bonferroni corrected importance degree of p = 0.00170 and p = zero.00005 in unsupervised PCA evaluation.

Breast cancer:

A cohort of 18 samples from breast-cancer tumors was analyzed via GCxGC-HRMS. The capabilities described allowed the analysis of the grades that become beneficial in discriminating between samples of various grades, as labelled with the aid of a most cancer pathologist. In addition, the approach described may offer advantages of the usage of smooth ionization with GCxGC, which would offer more direct characterization of the molecular composition.

Metabolic profiling

A team recently engineered a newly developed GC microwave derivatization method for the analysis of compounds of relevance to *metabolite profiling*, with GC-MS and GCxGC-ToFMS strategies used to display the derivatized merchandise. This changed into screening of in-born errors of metabolism (IEMs) in infant urine. The method substantially reduced the time taken for derivatization, reported more sensitivity, and produced fewer side reactions (a more entire derivatization reaction) when in comparison to conventional derivatization approach, as reviewed by Kouremenos et al, 2012 (31).

PREPARATIVE GAS CHROMATOGRAPHY

The term preparative GC (pGC) applies to the usage of large-diameter columns to purify relatively huge amounts of samples. This involves the use of a sample collection system at the column outlet. The pGC system consists of a carrier gas system, an injection system, a separation column, splitter and detector and a trap used to collect the effluents.

The sample fraction collection system that finds major use is Gerstel PFC system that contains 6 traps for collection of samples and one trap for waste. Following separation on a GC column, the system gathers the sample aliquots individually. The use of a microprocessor aids in automation.

The technique of preparative GC has been applied to several arenas of research such as:

The separation of volatile compounds from oils: for instance, 1,4-Dimethoxybenzene or Menthol and menthone from peppermint oil and β-Elemene, curzerene, curzerenone, curcumenol and curcumenone from *Curcuma rhizome*.

The separation of stereoisomer reaction products from manufacturing reactions to elucidate reaction mechanisms. For example, Isotope of nitrogen were separated from natural gas, Enantiomers of methyl 2-chloropropionate were separated from a racemic mix.

Purification of compounds that occur in trace levels such as Juvenile hormone III from a mix, Naphthalene, phenanthrene, fluoranthene, pyrene, chrysene and benzo[α]pyrene from aerosols, as reviewed by Zuo et al, 2013 (3).

CONCLUSION

Since its inception, GC has revolutionized the science industry due to its ability to separate and quantify samples in a mixture. The various aspects of instrumentation in a GC set up were discussed with the several types of carrier gases, columns, phases and detectors. Additionally, the necessity and reagents used in pretreatment of samples were elucidated in the section titled " derivatization". The several applications of GC: qualitative and quantitative were described with the help of case studies. The technique that was initially devised for the separation of volatile fatty acids has spread its roots to vast realms of science such as health, ecology, diagnosis and the separation of a huge repertoire of compounds.

REFERENCES

1. Martin AJP, Synge RLM. A new form of chromatogram employing two liquid phases: A theory of chromatography. 2. Application to the micro-determination of the higher monoamino-acids in proteins. Biochem J.1941; 35: 1358-1368
2. James AT, Martin AJP. Gas-liquid partition chromatography; the separation and micro-estimation of volatile fatty acids from formic acid to dodecanoic acid. Biochem J. 1952; 50: 679-690.
3. Zuo H-L, Feng-Qing Yang, Wei-Hua Huang, Zhi-Ning Xia; Preparative Gas Chromatography and Its Applications. *J Chromatogr Sci.* 2013; 51 (7): 704-715. doi: 10.1093/chromsci/bmt040
4. Rahman MM, Abd El-Aty A, Choi J-H, Shin H-C, Shin SC, Shim J-H. Basic Overview on Gas Chromatography Columns. Anal Separation Sci. 2015; 3:3:823–834.
5. Stashenko E, Martínez JR. *Gas Chromatography-Mass Spectrometry, Advances in Gas Chromatography*, Dr Xinghua Guo (Ed.), InTech. 2014; DOI: 10.5772/57492.
6. Pravallika S. Gas Chromatography a Mini Review. Research and Reviews J Pharm Anal. 2016; 5 (2): 55 62.
7. McNair HM, Miller JM. Basic Gas Chromatography, John Wiley & Sons, Inc. New York. 2009. Second Edition. ISBN: 978-0-470-43954-8.

8. Ettre LS, Golay MJE. the invention of open-tubular (capillary) columns. J High Resolut Chromatogr. 1987; 10: 221–230.
9. Jennings W. Evolution and application of the fused silica column. J High Resolut Chromatogr. 1990;3: 601–608.
10. Russo MV. Chiral separation of methoxamine and lobeline in capillary zone electrophoresis using ethylbenzene-deactivated fused-silica capillary columns and cyclodextrins as buffer additive. J Pharm Biomed Anal .2002;29(6):999–1003.
11. Ong RCY, Marriott PJ. A Review of Basic Concepts in Comprehensive
12. Two-Dimensional Gas Chromatography. Journal of Chromatographic Science. 2002; 40; 276-290.
13. Zellner Bd'A, Bicchi C, Dugo P, Rubiolo P, Dugo G, Mondello L. Linear retention indices in gas chromatographic analysis. Flavour Fragr J. 2008; 23: 297–314.
14. Orata F. Derivatization Reactions and Reagents for Gas Chromatography Analysis, Advanced Gas Chromatography - Progress in Agricultural, Biomedical and Industrial Applications, Dr. Mustafa Ali Mohd (Ed.). 2012; ISBN: 978-953-51-0298-4, InTech.
15. Pacholec F, Butler HT, Poole CF. Molten organic salt phase for gas–liquid chromatography. Anal Chem. 1982;54:1938–1941.
16. Pacholec F, Poole CF. Stationary phase properties of the organic molten salt ethylpyridinium bromide in gas chromatography. Chromatographia. 1983;17, 370–374.
17. Devic T, Serre S. (2009) Ordered Porous Solids: Recent Advances and Prospects (eds V. Valtchev, S. Mintova, and M. Tsapatsis), Elsevier, Amsterdam.
18. Chang N, Gu ZY, Wang HF, Yan XP. Metal-organic-framework-based tandem molecular sieves as a dual platform for selective microextraction and high-resolution gas chromatographic separation of n-alkanes in complex matrixes. Anal Chem. 2011;83, 7094–7101.
19. Ferey G, Mellot-Draznieks C, Serre C, Millange F, Dutour J, Surble S, Margiolaki I. A chromium terephthalate-based solid with unusually large pore volumes and surface area. Science. 2005;309, 2040–2042.

20. Valcarcel M, Cardenas S, Simonet BM. Role of carbon nanotubes in analytical science. Anal Chem. 2007;79, 4788–4797.
21. Agah M, Potkay JA, Lambertus G, Sacks R, Kensall D. High-performance temperature-programmed microfabricated gas chromatography columns. J Microelectromech. Syst. 2005;14 (5), 1039–1049.
22. Biemann K. Four Decades of Structure Determination by Mass Spectrometry: From Alkaloids to Heparin. J Am Soc Mass Spectrom. 2002;13,1254-1272.
23. Cardoso CAL, Vilegas W, Pozetti GL. Gas chromatographic analysis of indole alkaloids from Tabernaemontana hilariana. J Chromatogr A. 1997;788:204.
24. *Chromatography and Its Applications*: Edited by Sasikumar Dhanarasu. InTechOpen, 2012.
25. Groneman AF. Identification and determination of metabolites in plant cell biotechnology by gas chromatography and gas chromatography/mass spectrometry: Application to Nonpolar Products of Chrysanthemum cinerariaefolium and Tagetes Species. Analytica Chimica Acta. 1984;163:43-54.
26. Eiserbeck C, Nelson RK, Grice K, Curiale J, Reddy CM. Comparison of GC–MS, GC– MRM-MS, and GC×GC to characterise higher plant biomarkers in Tertiary oils and rock extracts. Geochim Cosmochim Acta. 2012; 87:299–322.
27. Frysinger GS, Gaines RB. Oil Spill Source Identification by Comprehensive Two-Dimensional Gas Chromatography (GC × GC) J Sep Sci. 2001;24:87.
28. Aguiar A, Silva Júnior AI, Azevedo DA, Aquino Neto FR. Identification of Methylhopane and Methylmoretane Series in Ceará Basin Oils, Brazil, Using Comprehensive Two-Dimensional Gas Chromatography Coupled to Time-of-Flight Mass Spectrometry Fuel. 2010; 89: 2760.
29. Kiepper, A.P, casilli A, Azevedo DA. Depositional paleoenvironment of Brazilian crude oils from unusual biomarkers revealed using comprehensive two dimensional gas chromatography coupled to time of flight mass spectrometry. Organic Geochemistry. 2014; 70:

62-75.

30. Tran TC, Logan GA, Grosjean E, Ryan D, Marriott PJ. Use of comprehensive two-dimensional gas chromatography/time-of-flight mass spectrometry for the characterization of biodegradation and unresolved complex mixtures in petroleum. Geochim. Cosmochim. Acta. 2010;74 :6468.

31. Kouremenos KA, Johansson M, Marriott PJ. Advances in Gas Chromatographic Methods for the Identification of Biomarkers in Cancer. J Cancer. 2012;3:404-420. doi:10.7150/jca.4956.

CHAPTER 6

GAS CHROMATOGRAPHY MASS SPECTROMETRY AND LIQUID CHROMATOGRAPHY MASS SPECTROMETRY

INTRODUCTION

As described earlier, conventional chromatographic methods, including gas chromatography was never been acknowledged to be the most suitable technique, to analyse components that are present in very low concentration in a particular mixture. In accordance with which, a search for an advanced version was initiated in order to measure specific organic volatile hazardous pollutants with more accuracy and higher precision, without any technical hassles.

In this regard, a combination of a gas chromatography and a mass spectrometry may appear as the obvious choice for identifying true components of the complex mixtures (1).

PRINCIPLE OF GC/MS INTERFACE

Gas chromatography coupled with mass spectrometry can be referred to as a very robust automation to identify, separate, quantify and analyse unknown volatile compounds in complex environmental mixtures. The technique is based on the principle that when heated, a particular mixture can be segregated into distinct analytes, which can be passed through a column along with a natural inert gas, like helium. Further to their separation, these analyte species can directly be analysed through mass spectrometry for their quantification (2)

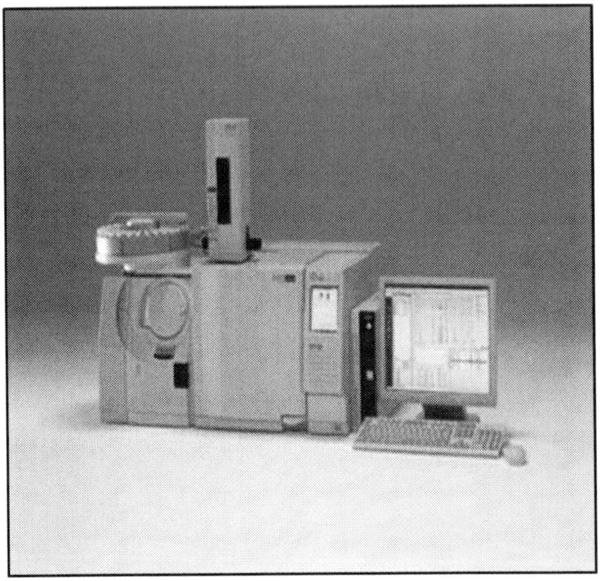

Instrumentation

Gas Chromatogram/Mass Spectrometer, capable of separating targeted analytes from mixture and identifying compounds within specific atomic mass unit. i.e. 45-300 amu. The system has integrated calibration system; for additional data quantification, a personal computer with GCMS software is needed.

The system is paired with the appropriate GCMS software, that enables to detect, analyse and quantify the targeted compounds; and the data is being stored for further analysis.

The instrument is equipped with the sampling probe that is made up of glass or stainless steel of appropriate length and material to sustain on & off heating and accidental adsorption of gas samples.

A ***leak proof, heated pump*** made up of KNF Neuberger or equivalent material is provided along with the analyser to maintain required sample flow rate.

Calibration assembly, an apparatus allowed to introduce calibrated gases through probe outlet into the sampling system, with the provision to monitor sample pressure, both during

calibration as well as sample passage. The assembly should also have a provision to monitor a temperature at the point of introduction.

Sampling Line is provided as a connector to pass sample to the GCMS analyser. The sample line is heated to offer appropriate temperature required to prevent sample condensation and adsorption of targeted compounds. The sampling line is made up of suitable low adsorbent stainless-steel material or Teflon material as specified by the regulation.

Sample condenser system, equipped with particular cooling system required to remove moisture content of the sample, to a level acceptable for sample injection system.

The analyzers are supported with *sample flow rotometers*, capable of withstanding sample gas conditioning.

Sample transfer line passages sample from its interfacing system to the inlet of GCMS instrumentation. The sample transfer line is generally heated to prevent unnecessary sample condensation. Particulate filter made up of 0.3-micron filtration capacity is placed in an appropriate position to remove moisture content of the gases.

Calibration gas are provided for their easy delivery through the sampling system. The calibration gas flow rate is being measured and controlled through the introduction of mass flow meters or controllers. Teflon tubing with suitable length and diameter to connect with regulators of cylinders can be attached to the analyzer.

Data Analysis

The samples can be quantified with the formula specified below, to measure the concentration of targeted VOHAP's in a given gas mixture.

ppmv Target VOHAP = (Ax/ARRF) (Cis/Ais)

Where

A_x = Peak area of target VOHAP quantion.

A_{is} = Peak area of corresponding selected internal standard quantion.

C_{is} = Concentration of corresponding internal standards

ARRF = Average Relative Response factor of target VOHAP calculated from three-point calibration.

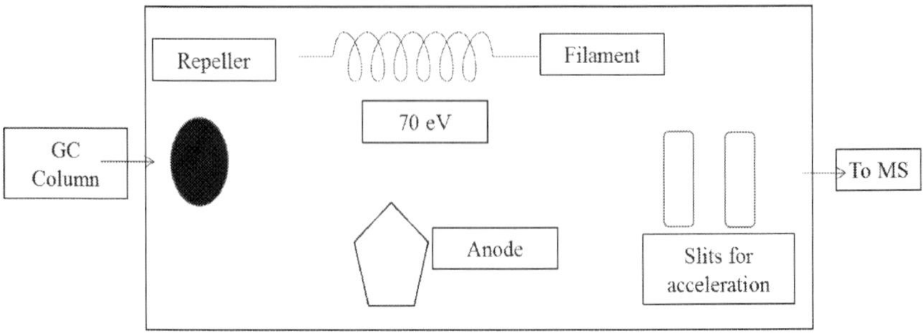

Figure 6.1: A schematic of GCMS workflow

For accurate outcome, the column outlet needs to be directly attached to the mass spectrometer, through its ion source. Moreover, different strategies that are ought to be implemented, should pass below mentioned check points.

- A condensation of the analyte in the interface should be avoided.
- Analyte should not decompose, within the interface or before entering into ion source.
- The flow of gases, as regulated by the mobile phase gas flow rate, and should be kept within its pumping capacity.

SCOPE AND APPLICATIONS OF GC-MS

Direct interface of GC with mass spectrometry

The method deals with a direct interfacing of gas chromatography with mass spectrometry, specifically designed for hassle free identification, separation and quantification of mainly 36 different volatile organic compounds, present in the environmental samples. Although, these 36 volatile compounds can be referred from a chart mentioned herewith; the scope is not limited to them and can be extended to many other compounds.

Determination of chemical volatile compounds

The method is believed to be suitable for the determination of above mentioned chemical volatile compounds, while emitting from their stationary sources (3). Moreover, some of the hazardous air pollutants, which are volatile in nature (VOHAPs) can as well be identified, separated and quantified by direct interfacing of gas chromatography and mass spectrometry measurement system, to the source effluent.

Table 6.1: List of a few volatile compounds that can be determined by GCMS (3,4)

Sr. No.	Compound	Sr. No.	Compound
01	Benzene	19	Cis-1,2-dichloroethene
02	Bromodichloromethane	20	Dibromochloromethane
03	Carbon Disulphide	21	1,1-Dichloroethane
04	Chloroform	22	1,2-Dichloropropane
05	Methyl Isobutyl Ketone	23	Ethyl Benzene
06	Styrene	24	Ethyl Chloride
07	Tetrachloroethylene	25	1,1,2,2-Tetrachloroethane
08	Toluene	26	1,1,2-Trichloroethane
09	Bromoform	27	Bromomethane
10	Vinyl Acetate	28	Chlorobenzene
11	Vinyl Chloride	29	1,2-Dichloroethane
12	Chloromethane	30	Methyl Ethyl Ketone
13	1,1-Dichloroethene	31	t-1,2-Dichloropropene
14	Methylene Chloride	32	m-Xylene
15	1,1,1-Trichloroethane	33	2-Hexanone
16	p-Xylene	34	Trichloroethene
17	Carbon Tetrachloride	35	0-Xylene
18	c-1,2-Dichloropropene	36	t-1,2-Dichloroethene

Direct measurement of moisture content of an unconditioned sample

The method can be applied for directly measuring the moisture content of an unconditioned sample that is less than its saturation value at applicable instrument operating limit (2-4). However, samples with more than specified limit of moisture content may need conditioning before infusing into the analytical instrument to prevent further application error, due to moisture condensation within the instrument; for that matter additional QA has been provided for analysis of polar and water-soluble compounds.

RANGE AND SENSITIVITY OF THE METHOD

From a data collected so far, the measurement range of the instrument can be defined within 150 ppbv to 100ppmv; however, samples with concentration outside this range can be measured in case:

- The performance pre-requisites of the method are fulfilled.
- Concentration used to prepare the three points are adjusted appropriately with calibration concentrations.
- The three-point calibration is extrapolated to include additional operations.

The ***sensitivity*** of GC-MS measurement system towards particular target analyte mainly depends upon:

- The specific instrumental response for each target analyte
- Number of quantified mass spectrum ions available.
- The amount of instrument noise and the moisture content of the sample.

Objectives for achieving accurate data quality

One of the primary objectives for perfect data quality is to achieve an accuracy between the range of +/- 20% and a precision between the range of +/- 10%; for each measured value. This objective is mainly fulfilled to identify compounds that are not been calibrated through three-point process and to facilitate comparative analysis with other mass spectrometer.

A continuous hassle-free operation is carried out with maximum accuracy and precision, by establishing three-point calibration reference data for standardized MS tune condition, GC operating conditions and data quantification; which can be referred while acquiring and quantification of analyte. In addition, proper preventive maintenance and corrective action techniques are implemented, while performing effluent measurements. Finally, accurate data quality would avoid unacceptable results due to loss of water soluble compounds and/or polar compounds, due to moisturization; through implementation of proper corrective action.

CASE STUDY: A BRIEF SUMMARY OF A METHOD

Principle

The automated version of instrument is identified to be a versatile tool to disintegrate gaseous mixtures of molecules from each other, on the basis of their affinity with column's stationary as well as mobile phase. An internal standard mixture must be quantitatively co-added with every sample, in order to avoid variation of results. A compound that is to be analyzed with this technique has to be sufficiently volatile in nature and should thermally be stable.

Operations

Sample gas can be injected into the GC inlet via pneumatic valve assembly; and is being carried out by a carrier gas, which generally is helium through chromatographic column. On the basis of relative affinity of each compound towards the GC column, these compounds get separated from the mobile phase gas carrier and further enter mass spectrometer. Since operating ambient pressure of the mass spectrometer is lower than that of the gas chromatography, an interface is suggested to pass on the compound from gas chromatography to the mass spectrometer (Figure 6.2).

Thus, the compound is passed through a heated transfer valve, and is subjected to ionizing energy; causing ejection of an electron from the molecule. As a result of which, the compound is being converted into many positively charged ion fragments and stable electronic configuration is being achieved.

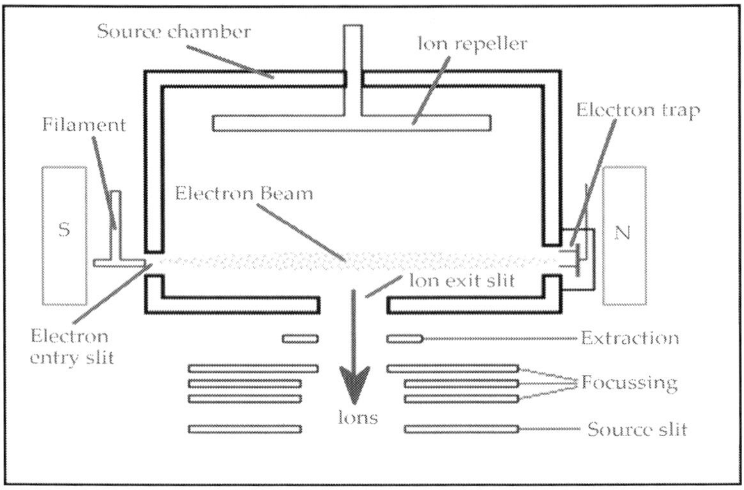

Figure 6.2: Schematic of GCMS instrument for the determination of volatile compounds

Once the compound is being scanned by the mass spectrometer, it detects those positively charged unique fragments, within its defined mass range (45-300 amu). These fragments are positively charged ions or radical cations, produced due to large amount of energy imparted during loss of electron; either in a process of electronic ionization or chemical ionization. The characteristic relative abundance of these radical cations can offer structural fingerprints of particular compound of interest.

These positively charged ions are being passed through a mass analyzer and are being separated on the basis of mass related properties. However, it should be noted that this separation can be variable depending upon the type of analyzer used. From different types of analyzers that are currently into existence, such as quadrupoles, ion traps, magnetic sector, time-of-flight, radio frequency, cyclotron resonance, etc.; the quadruples and ion traps are being focused more popularly.

The separated ions are then channelized through a detector system; wherein the signal is being amplified depending upon their numbers to provide a peak. The data is being further sent to a computer, which decodes those electrical impulses into a visual display for further analysis (5).

Thus, mass spectrometer analyses the target compound through:

- Comparison of GC peak retention time in the total ion chromatogram, for eluting compounds with that of the internal standards in the three-point calibration.
- Through examination of particular mass spectral fingerprints of the peaks for eluted compounds.

Sampling

A sample interface system should be calibrated in order to provide samples to the GCMS analysers in their equilibrated state. Moreover, it should as well include additional provision to facilitate inbuild instrument and system calibration; when the gases are being introduced at the extrapolative probe outlet of the system. A continuous operation is another pre-requisite for the sample interface system to avoid processing delay between consecutive samples.

The ideal ***sample interface system*** comprises of a heated particulate filter, attached to a heated pump and Teflon sample line; as well as a sample conditioning unit. For continuous operations, the sample can be extracted from the duct and passed through the interface at a constant and a steady rate. The temperature plays a very important role to avoid unnecessary sample condensation and to allow error free operations.

The primary objective of the conditioning unit is to prevent sample loss due to moisture imbalance and condensation and to remove excess moisture that may interfere with the detection system. In order to provide an acceptable arrangement, following precautions should be implemented:

- Except for the condenser, all other component within the conditioning unit should be maintained at higher temperature, comparable to that of introduction valve assembly.
- The secondary particulate filter should be maintained at a temperature same as that of the GC introduction valve assembly.

The sample line, is denoted as a heated connection that is responsible for transfer of sample gas to the GCMS analyzer, at a substantially lowered rate. A special care is to be taken with the excessive sample gas, which should be vented at an atmospheric pressure to avoid extra pressure on the sample transfer line.

Interferences

Although, the technique is considered to be a very definitive approach for confirming and separating different compounds in a particular mixture, targeted GC/MS system does have certain limitations, which are to be overcome, for error free operations and applications, such as:

Analytical Interference

It is considered to be those interferences, due to contaminating analytes that results in the false positive production of chromatographic peak and quantionic overlap; thus, interfering with the quantification of a compound of an interest. Moreover, two compounds of the same mass spectrum cannot be quantified by this system; without optimum conditions that allow adequate separation of the compound from the mixture, prior to analysis. Additionally, in case of optical isomers, an installation of special column is needed for exact separation and quantification of the same.

Sampling System Interferences

The interference of hazardous sampling system, like water, reactive particulate matter, adsorptive lines, and acidic gases can prevent easy transportation of targeted compound to the GC analyzer; or may as well impart potential damage to the measurement system components.

Precautions

It is to be noted that many of the target analytes are toxic in nature and are known carcinogens. Thus, precaution is to be taken for limited chemical exposure, while handling them. These mixtures are compressed and packed in gas cylinders, therefor proper precautionary measures, should be incorporated, while their transportation and handling. While choosing an appropriate location for sampling, it is advisable to avoid high positive or high negative pressure. A controlled environment needs to be provided for optimum temperature, and pressure requirements. It should as well be noted that in order to avoid accidental exposure and transport hazards, a regular leak check of the sampling system should be carried out, along with the inspection and maintenance of sample exhaust equipment (2, 4,5).

LIMITATIONS AND CONCERNS OF GC-MS

When done a comparable cost analysis, it has been acknowledged that costly raw materials and accessories, like ion sources, mass filters, pumps, metal detectors and hardware as well as the software instrumentation, may limit its accessibility and wide range of applications.

However, the entire technique is being backed up due to accuracy that has been offered in exact determination of the analyte, instead of just a qualitative signal detection, which may vary with the amount of the analyte.

Moreover, sample analysis can be performed with a longer retention time, but the said concern may be offset, due to upgraded versions of portable GC/MS models.

LIQUID CHROMATOGRAPHY-MASS SPECTROMETRY (LC-MS)

The combination of Liquid Chromatography-Mass Spectrometry (LC-MS) refers to an analytical technique that amalgamates both separation with high resolving power, using chromatography as well as sensitive detection using mass spectrometry. The Liquid chromatography involved may entail the following types: High Performance Liquid Chromatography (HPLC)-MS, Capillary Electrophoresis (CE)-MS and Capillary Electrochromatography (CEC)-MS (5). The technique of LC-MS has now a large area of application with several tests conducted using the platform ranging from toxicity studies to screening and hormone studies (6).

PRINCIPLE OF LC-MS

The technique of LC-MS is typically a HPLC-MS combination along with an ionization source for generation of sample fragments. A sample containing analytes is first subjected to separation on a LC column following which the resolved samples are subjected to ionization.

This generates samples in the form of ions in the gas phase. These are passed into a mass analyzer that separates the ions on the basis of mass to charge ratio. A detector then identifies the separated ions exiting the mass analyzer followed by generation of a signal specific for an ion. A graphical representation of a plot of the signal of an ion species against its mass to charge ratio is generated called mass spectrum that yields data about a molecule, its structure and mass (7,8).

INSTRUMENTATION

Usually the technique uses a LC instrument composed of:

- A pump.
- A system for removing gas from sample and mixing of solvent
- A sample injector that can be automated or manual
- The analytical column (s)
- Detector and Recorder

The detectors used may be either UV based (single or variable or multiple wavelengths) or Florescence based or an electrochemical sensor.

REQUIREMENTS OF LC-MS INSTRUMENTATION

The system is composed of the apparatus for liquid chromatography that is linked to a source of ionization, a mass analyzer and a data acquisition unit.

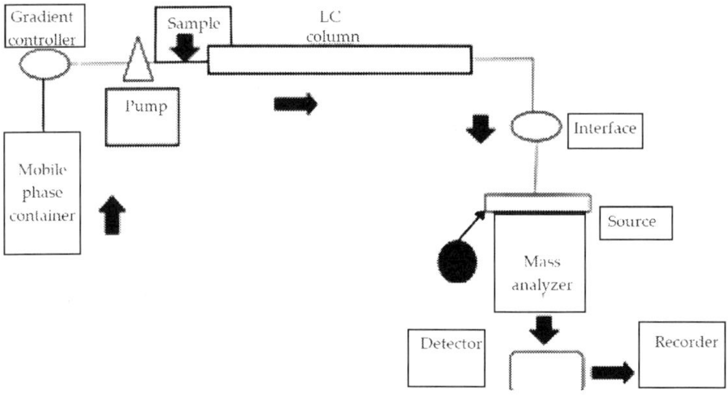

Figure 6.3: The work-flow of LC-MS

There is an Interface that connects the outlet from the liquid chromatography unit along with the ionizing unit of the mass spectrometer

SOURCES OF IONIZATION

Ionization involves a process of addition or removal of electrons from a sample to produce ions, in the case of LC-MS the need of vapor or condensed phases is necessitated (9).

There are several techniques of ionization that include: Atmospheric pressure ionization (API), Electrospray ionization (ESI) and Matrix-assisted laser desorption/ionization (MALDI).

Atmospheric Pressure Ionization

The technique involves the simultaneous elimination of solvent molecules and ionization. The solvent molecules used in the LC process are removed and the gas analytes thus produced are subjected to ionization. There are two main methods of ionization:

Electrospray ionization (ESI)

The application of a a strong electric field on a liquid that traverses through a capillary tube generates an electrospray to produce

Electrospray is produced by applying a strong electric field to a liquid passing through a capillary tube with a weak flux. This produces charged large droplets which is then subjected to evaporation. The removal of droplets of solvent from the analytes results in coulombic repulsion forces to balance the surface tension leading to the formation of ions from the droplets.

The technique has a sensitivity of 10^{-8} μl and enables the ionization of samples that have a large mass and are non-volatile. However, the challenges with the technique are instability, a reduction in sensitivity as well as lesser fragmentation (9,10).

Atmospheric pressure chemical ionization (APCI)

Similar to ESI, the samples enter a capillary tube but instead of an electric field, the technique is based on nebulising the mobile phase with an inert gas. Following interaction with the inert gas it is vaporized by heating to elevated temperatures (greater than 673K) following a discharge electrode to form ions. The most ubiquitous method for ionization is the APCI for the analysis of a host of lab or commercial or clinical analytes (5).

Atmospheric pressure photo ionization (APPI)

The technique has been developed for samples of low polarity such as polycyclic compounds that are not amenable to ESI or APCI. An aerosol of samples is exposed to radiation with light to cause the emission of photons to form ions (Crawford Scientific)

Matrix-assisted laser desorption/ionization (MALDI)

This technique is applied to huge samples that are sensitive such as polymers, proteins, peptides, fullerenes and dendrimers. The analytes are embedded in a matrix that is irradiated with laser (The matrix absorbs the laser wavelength): ultraviolet (UV) lasers (337 nm) have been reported to be applied on the surface to cause the production of ions. The mechanisms behind the ionization involve the quasithermal evaporation of particles of the matrix upon contact with laser to increase the motion of molecules. This leads to desorption of the top layers of matrix that transfers its protons to snalytes to ionize them.

The matrix is responsible for absorption of energy from the laser and transferring it to the solid support and also dissolve the analyte to minimize the forces and interactions between the sample molecules. A protocol described for MALDI includes:

Dried-droplet method: A matrix sample ratio of 5000:1 is applied as a 0.5–2.0 μL drop on the analyte and dried.

Thin layer method: The sample is layered on the matrix for its subsequent absorption. This technique has been reported as sensitive and accurate.

Sandwich method: Matrix crystals are prepared thinly to be added to trifluroacetic acid and samples and the matrix again (5)

ANALYZER

Mass spectrometer analyzer

The separation of ionized molecules is performed by the analyzer on the basis of charge to mass ratios to pass them on for subsequent detection. There are several types of analyzers that include:

Quadrupole Mass analyzer

The analyzer utilizes electric field as sample ions pass a central axis surrounded by four parallel poles that are at equal distances. This Quadrupole has each pair of metal rods facing each other through which voltage of Radio Frequency is applied. Depending on the voltage, ions of a requisite mass-to-charge traverse and are detected while ions that follow an unstable trajectory are disintegrated. The technique has a high resolution coupled with accurate masses and good speed with linear outputs that are used in quantitative studies.

Ion trap analyzer

The technique uses a trap: a ring electrode and two end cap electrodes to capture ions. Following application of voltages, the ions of specified mass to charge ratios can be trapped on the device, dependent on the voltage applied. The ions out of range traverse trajectories that make them leave the analyzer. The technique in conjunction with an inert gas yields fragments for studies on structure.

Time of Flight (TOF) analyzer

The ions produced are exposed to an increasing voltage to traverse an evacuated tube of known dimensions. Once the voltage is released, samples separate at a rate dependent on their mass to charge ratios. The technique is rapid and sensitive.

Fourier transform-ion cyclotron resonance (FT-ICR)

The ions generated enter a chamber subjected to electrical and magnetic fields that causes them to enter circular orbits. Excitation by an electrical field causes the production of a current per time that is transformed into orbital frequencies by Fourier transform which in turn is dependent on the mass-to-charge ratios. These analyzers can analyze a wide variety of mass with high resolution.

Magnetic Sector Mass Analyzer

The ions generated enter a flight tube to separate on the basis of charge to mass ratio by application of a magnetic field. Depending on the ration ions that are deflected to the similar extent follow a trajectory while others are unable to enter the detector (5, 7, 8).

DETECTORS

Following the analysis and separation, the sample analytes require detection and quantification into a signal. The detectors can either be point detectors that involve striking of ions one after another at a single point or array detectors that detect ions simultaneously using several detection points.

There are three detectors applied that include:

- Electron multipliers that convert ions of any charge into electrons that are amplified and detected.
- Dynolyte photomultipliers effect a conversion of sample charges into electrons that interact with a phosphor material. This leads to emission of photons to be detected by a photomultiplier.
- Microchannel Plates (MCP) are commonly used along with ToF spectrometers. Response times have been reported as very low less than a nanosecond.

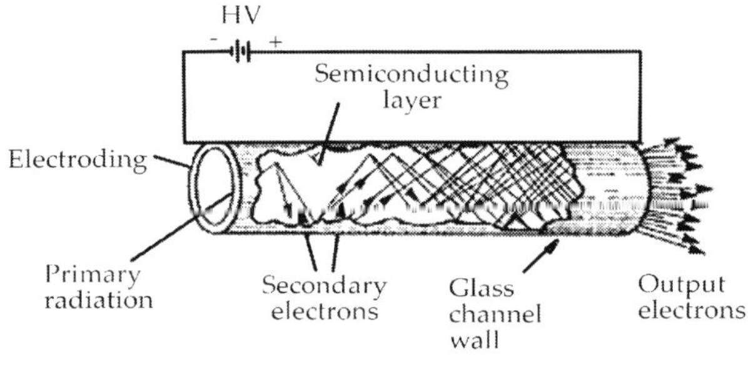

Figure 6.4: The schematic of an electron multiplier (courtesy NASA)

APPLICATIONS OF LC-MS

Clinical applications:

The technique of LC-MS has seen a boost in its applications especially in labs due to the following advantages:

- Molecules with low molecular weight were assayed by competitive radio-immunoassays (RIAs) that entail the use of radioactivity for detection and an additional issue of limited range.
- As an alternative, HPLC and GC-MS were used that have a lower throughput than compared to LC-MS.
- The technique is more economical than other assays for testing. (14)

CASE STUDIES

Detection of Thyroglobulin

Thyroglobulin has been reported as a marker for thyroid cancer patients. Due to the presence of antibodies in the plasma of 1/5 th of these patients, the analysis of this marker is challenging. A 2008 study developed LC-LC-MS/MS in order to measure the levels of thyroglobulin in patients who possess anti-thyroglobulin autoantibodies (15).

Detection of Vitamin D

The hydroxylation of Vitamin D yields 25-hydroxy vitamin D_2/ D_3 (25OHD) that circulates in the blood. The measurement of this form has been achieved by several techniques such as HPLC or chemiluminescence. These techniques have yielded a large amount of variations in the results. As a solution, National Institute of Standards and Technology characterized by LC-MS-MS a human serum that contains the molecule and also enables the precise quantification of the technique (14).

Detection of Parathyroid hormone

The elevated levels of calcium, cancers are associated with levels of Parathyroid hormone that is present at very low levels and also is detected with cross reactivity in commercial assays. Following sample clean up a research team detected the level of parathyroid hormone using LC/MS-MS and reported lesser levels of cross reactivity as compared to commercial tests (14).

Entomotoxicology

This involves the study of drugs in insect systems. The field of medico-legal autopsies involves the use of insect larvae for calculation of postmortem interval. Insect systems can serve as systems for study of toxicology. The field of forensic entomology is being explored by Entomology of the Royal Belgian Institute of Natural Sciences and the Toxicology and Microtraces sections of the National Institute of Criminalistics in Belgium along with Academic Hospital of the Free University of Brussels. For instance, the technique of LC–MS–MS for the detection of Nordiazepam and its metabolite Oxazepam in *Calliphora vicina* successfully (16).

Textiles

An important part of forensic fibre examination involves the characterization of textile dyestuffs. Textile fibres found at a crime scene can be used as chemical evidence in a wide range of crimes. Current methods are not sufficient for unambiguous identification because similar dyes having the same colour might have different molecular structures. Projects are on-going to characterize and compare fibres according to molecular structure of the extracted dye. A reference database of LC–MS spectra of these dyes will be created (16).

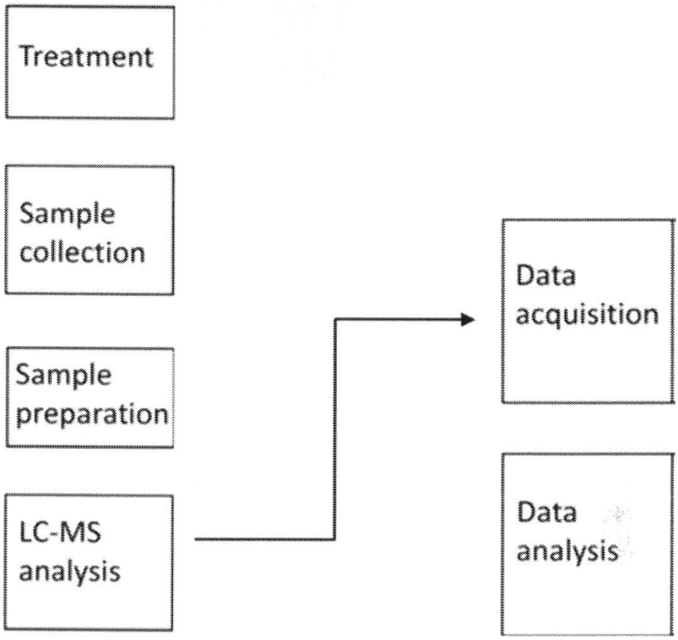

Figure 6.5: Schematics for LC-MS analysis of metabolomics

Doping tests

The International Olympic Committee (IOC) still relies on capillary GC–MS and/or high-resolution mass spectrometry for confirmation of positive findings. The use and abuse of anabolic steroids is a growing problem. Current detection techniques suffer from an extensive sample pretreatment and thus from low sample throughput. The European Commission started the SGLC/MS Project: This project involves the development of new test methods that will allow modern drug testing techniques to be more widely and effectively used. The preparation of suitable reference compounds is likely to play a vital role in this process (16).

Corticosteroids and stimulants are the class of compounds that are illicitly used by professionals owing to their anti-inflammatory and mood elevating as well as euphoric properties respectively. They are widespread abused among sports persons. Counterfeiting of herbal drugs with synthetic agents can also be detected by LC–MS/MS. Different mass analyzers (TOF, IT (ion trap)) have been coupled with LC and extensively used for screening and characterization of different analytes in plasma and herbal extracts (17).

LC-MS and determination of structures

The uniqueness of LC/MS is the combined high resolving power of HPLC and superior mass detection capability of MS. The LC/MS and LC/MS/MS techniques provide solutions to a wide range of structural characterization problems in pharmaceutical research. They include identification of trace level impurities and degradants in bulk drug substances, identification of metabolites in drug metabolism studies as well as unknown identifications in synthetic reaction product. The uses of LC/MS and LC/MS/MS techniques can rapidly provide structural information of unknown impurities found in production batches. Another application of the LC/MS and LC/MS/MS technique is the identification of metabolites including reactive species in drug metabolism studies. Looking at metabolomics, proteomics and biomarker discovery, it is expected that LC/MS technique will continue to play important roles in every aspect of drug discovery and development (18).

REFERENCE

1. Braithwaite A, Smith FJ. Chromatographic Methods. Blackie Academic & and Professional. PP 375-379. UK 1996.
2. Alltech GC/MS Switching Valve. Application note. Copyright 2004 Alltech Associates, Inc.
3. http://www.scientific.org/tutorials/articles/gcms.html
4. https://www.epa.gov/sites/production/files/2015-12/documents/8260b.pdf
5. Parasuraman S, Anish R, Balamurugan S, Muralidharan S, Kumar KJ, Vijayan V. An Overview of Liquid Chromatography-Mass Spectroscopy Instrumentation. 2014; 5(2). 47-55.
6. Want EJ, Cravatt BF, Siuzdak G. The expanding role of mass spectrometry in metabolite profiling and characterization. Chembiochem. 2005;6:1941–51.

7. Korfmacher WA. Foundation review: Principles and applications of LC-MS in new drug discovery. Drug Discov Today 2005; 10(20): 1357-67.
8. Lim CK, Lord G. Current development in LC-MS for pharmaceutical analysis. Bio Pharm Bull. 2002; 25(5): 547-57.
9. Ho CS, Lam CW, Chan MH, Cheung RC, Law LK, Lit LC, *et al.* Electrospray ionisation mass spectrometry: principles and clinical applications. Clin Biochem Rev. 2003; 24(1): 3-12.
10. Fenn JB, Mann M, Meng CK, Wong SF, Whitehouse CM. Electrospray ionization for mass spectrometry of large biomolecules. Science 1989; 246(4926): 64-71.
11. Crawford Scientific. Chrome academy pdf.
12. LC/MS booklet. Available in http://ccc.chem.pitt.edu/wipf/Waters%20 LC-MS%20primer.pdf
13. Mass Spectrometry (MS) Primer. Available in http://www.chemweb.ucc.ie/ Mass%20Spec/MS%20summary.
14. Grebe SK, Singh RJ. LC-MS/MS in the Clinical Laboratory – Where to From Here? The Clinical Biochemist Reviews. 2011;32(1):5-31.
15. Hoofnagle AN, Becker JO, Wener MH, Heinecke JW. Quantification of thyroglobulin, a low-abundance serum protein, by immunoaffinity peptide enrichment and tandem mass spectrometry. Clin Chem. 2008;54:1796–1804.
16. De Boeck G, Wood M, Samyn N. Recent Applications In LC–MS in Foresic Science. November 2002 Recent Applications In LC–MS. November 2002.
17. Haneef J, Shaharyar M, Husain A, Rashid M, Mishra R, Parveen S, Ahmed N, Pal M, Kumar D. Application of LC–MS/MS for quantitative analysis of glucocorticoids and stimulants in biological fluids. J Pharma Analysis. 2013; 3(5): 341-348.
18. Chen G, Pramanik BN, Yan-Hui Liu and Urooj A. Mirza. Applications of LC/MS in structure identifications of small molecules and proteins in drug discovery. J Mass Spectrom. 2007; 42: 279–287.

CHAPTER 7

HIGH PERFORMANCE LIQUID CHROMATOGRAPHY

INTRODUCTION

High Performance Liquid Chromatography is an analytical technique used for the separation, identification and quantification of each constituent present in a mixture. HPLC is an advanced technique of column liquid chromatography. In basic liquid chromatography, the solvent in the column usually flows with the help of gravity whereas in HPLC the solvent is forced under high pressure i.e. 400 atmospheres because of which sample get separated into different constituents on the basis of difference in their relative affinities (1, 2, 3, 4). Relative affinity means the tendency of a molecule or a substance to get associated with one another.

In HPLC, pumps will be used to pass pressurized liquid solvent including the sample mixture which is then allowed to enter into a column that is filled with solid adsorbent material. The interaction varies of each component and this variation causes difference in flow of each component which ultimately leads to separation of components of column.

HPLC mainly utilizes a column that holds packing material which act as stationary phase, a pump that moves the mobile phase through the column, and a detector that shows the retention times of the molecules. Retention time is a measure of the time taken for a salute to pass through a column. It is also calculated as the time from injection to detection. Retention time varies depending on the interactions between the stationary phase, the molecules being analyzed and the solvent used for separation (3, 4).

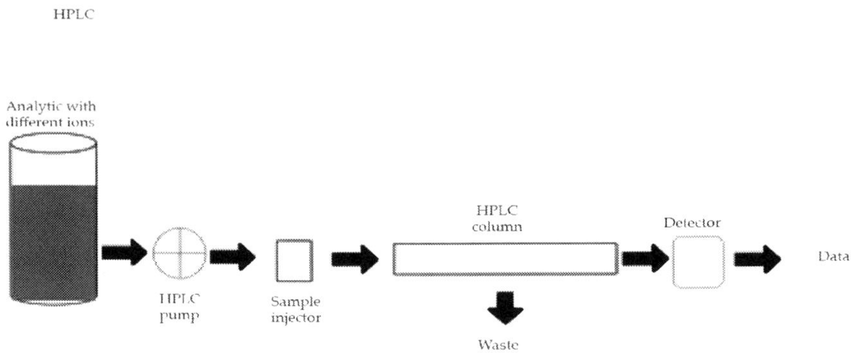

Figure 7.1: Schematic of HPLC elution

In other words, HPLC involves injection of a small volume of liquid sample into a tube packed with tiny particles (3 to 5 microns in diameter called stationary phase) where individual components of the sample are moved down the packed tube (column) with a liquid (mobile phase) forced through the column by high pressure delivered by a pump. These components are separated from one another by the column packing that involves various chemical and physical interactions between their molecules and the packing particles. These separated components are detected at the exit of this tube (column) by a flow-through device called detector that measures their amount. An output from this detector is called a liquid chromatogram.

PRINCIPLE OF HPLC

The basic principle behind HPLC process involve adsorption. HPLC relies on pumps to pass a pressurized liquid and a sample mixture through a column filled with adsorbent, leading to the separation of the sample components. The active component of the column, the adsorbent, is typically a granular material made of solid particles (*e.g.* silica, polymers, etc.), 2–50 μm in size. The components of the sample mixture are separated from each other due to their different degrees of interaction with the adsorbent particles. The pressurized liquid is typically a mixture of solvents (*e.g.* water, acetonitrile and/or methanol) and is referred to as a "mobile phase". Its composition and temperature play a major role in the separation process by influencing the interactions taking place between sample components and adsorbent. These interactions are physical in nature, such as hydrophobic (dispersive), dipole–dipole and ionic, most often a combination (2, 3, 6).

History of HPLC

In 1966, Yale University researchers published a paper on ion-exchange separation of organic compounds. A year later, they published on fast liquid chromatography using a high-pressure pump. In 1967, Waters Associates began marketing the first commercial HPLC instrument.

C&E cites 1969 as a landmark year for HPLC. Prior to that year, gas chromatography was the only separation analysis method for a variety of applications. A series of meetings in Las Vegas sparked interest in the then-emerging technology as an alternative to GC, which only worked for volatile compounds or ones that had been derivatized to become volatile. HPLC was capable of analyzing a broader range of compounds, making it more valuable in applications such as drug discovery and biological separations.

The first 6,000-psi HPLC devices were introduced in 1972, and the addition of septumless injectors evolved the "P" in HPLC from "pressure" to "performance." During the 1980s, chromatographers developed new separation methods and detectors to improve and expand the usability of HPLC into a variety of applications.

The 21st century has seen still more evolution in HPLC. New systems contain extremely small particles and higher pressures (15,000-20,000 psi). Faster analysis has long been a primary objective for HPLC development. Multi-dimensional HPLC addresses that by using two or more columns in sequence. Two-dimensional HPLC combines two chromatography modes — such as size exclusion and reverse phase — in which the output from one column feeds into a second. Each column has particles designed to separate different components.

Every day, HPLC helps scientists answer important questions about everything from food and beverage to medicines, household cleaners, pesticides and more. At present, HPLC is going strong and getting better with every passing year.

INSTRUMENTATION OF HPLC

The HPLC instrumentation involves pump, injector, column, detector, integrator, and display system. In the column, the separation occurs. The parts include:

Solvent Reservoir: The contents of mobile phase are present in glass container. In HPLC, the mobile phase or solvent is a mixture of polar and non-polar liquid components. Depending on the composition of sample, the polar and non-polar solvents will vary.

Pump: The pump suctions the mobile phase from solvent reservoir and forces it to column and then passes to detector. 42000KPa is the operating pressure of the pump. This operating pressure depends on column dimensions, particle size, flow rate and composition of mobile phase.

Sample Injector: The injector can be a solitary infusion or a computerized infusion framework. An injector for a HPLC framework should give infusion of the fluid specimen inside the scope of 0.1ml to 100ml of volume with high reproducibility and under high pressure (upto 4000 psi).

Columns: Columns are typically made of cleaned stainless steel, which are around 50mm to 300mm long and have an inward distance across, which is around 2 to 5mm. They are generally loaded with a stationary phase with a molecule size of 3μm to 10μm. Columns with inner diameters of <2mm are regularly alluded to as a microbore segments. The temperature of the mobile phase and the column should be kept consistent during investigation.

Detector: The HPLC detector, which is situated toward the end of the column, distinguishes the analytes as they elute from the chromatographic column. Regularly utilized detectors are UV-spectroscopy, fluorescence, mass- spectrometric and electrochemical identifiers.

Data Collection Devices or Integrator: Signals from the detector gathered either on graph recorders or electronic integrators that fluctuate in many-sided quality and in their capacity to process, store and reprocess chromatographic information. The PC coordinates the reaction of the indicator to every part and places it into a chromatograph that is difficult to interpret.

The schematic representation of a HPLC instrument ordinarily incorporates a sampler, pumps, and a locator. The sampler brings the sample into the mobile phase stream which conveys it into the column. The pumps convey the mobile phase through the column. The detector generates a sign relative to the measure of sample component rising up out of the segment. A computerized microchip and software control the HPLC instrument and give information data. A few models of mechanical pumps in a HPLC instrument that can combine various solvents in proportions to changing in time, producing a synthesis slope in the portable stage. Most HPLC instruments similarly have a column broiler that recognize altering the temperature at which the partition is performed (5, 6, 7, 8).

RECENT ADVANCES IN HPLC INSTRUMENTATION

Column switching:

Usually an HPLC system consists of a simple flow-through sequence of one or a few columns. If this sequence is branched and switching devices are employed to interface individual columns, it then becomes possible to exert powerful control on the quality and the basis for the separation and LC becomes a versatile and powerful separation system. This leads to the idea of column switching, which means that the carrier flow is changed by standard six-port high pressure valves, so a fraction of or all of the effluent from any primary column is selectively transferred to any one of a number of secondary columns, that form a network, for further separation. Switching within the network may be manual or automatic. The objectives of column switching are to:

- increase the chromatographic resolution and selectivity;
- enrich trace amounts of sample;

- protect sensitive detectors from contamination by co-extractives;
- prevent destabilization of the chromatographic equilibrium of the column by co-extractives; and
- to achieve further objectives or a combination of several objectives within one chromatographic network (9,10,11).

Three commonly used switching functions are defined, in order to optimize chromatographic parameters:

- *Column back flushing* is used to remove sample components that are strongly retained on the primary column and speeds up the analysis of complex mixtures without the need of gradient elution. After the fraction has eluted from the primary column and has been transferred for further separation on to the secondary column, this technique reverses the flow of the primary column to the waste.
- *Column selection* (or column bypass) refers to column switching systems (12,13) which are used to reduce long analysis times of analytes with very different polarities, using two or more (different) columns. Late eluting components are directed after the primary (short) column to the detector, whereas the fast eluting components are directed to the secondary (long) column, for further separation, before they are detected.
- *Recycling chromatography*, in comparison with the column selection technique, offers the possibility of increasing the number of theoretical plates by repeated use of the same column. A portion of the chromatographic effluent is re-injected onto the same column.

MULTI-DIMENSIONAL LIQUID CHROMATOGRAPHY

One-dimensional chromatography is widely applied to the analysis of real world samples in several fields. However, it does not provide sufficient resolving power for the separation of complex mixtures encountered in food analysis, environmental analysis, and identification of microorganisms, molecular biology, proteomics and metabolomics.

A solution to this problem is to develop ***multi-dimensional separation schemes*** where the dimensions are based on different separation mechanisms (14,15).

Two-dimensional (2D) LC:

2DLC can be theoretically used in several multi-dimensional combinations, offering increased peak capacity, selectivity and resolution, especially in the comprehensive LC mode (14 - 16,18).

2D-LC can be performed either off-line or in an on-line mode depending on the way transfers from the primary column effluent to the second column are established. Off-line 2D-LC is the most popular approach since it is easier and very simple. Fractions of the first-dimension effluent are collected manually or via a fraction collector, after which they are re-injected on the secondary column. However, the off-line approach can be time consuming, difficult to automate, non-reproducible and susceptible to sample loss and contamination. The on-line approach is faster and more reproducible, but it needs interfaces and is more difficult to operate (14,16).

A 2D separation is orthogonal if the two separation mechanisms are independent of each other and have different retention profiles or in other words, provide different selectivities. The two mechanisms must spread the sample components following two different retention patterns, over as broad a range possible for retention factors, so that broad separation spaces be used in both separation dimensions. The retention parameters employed in the second dimension must separate the components that the mechanism used in the first column elutes closely (15,17-19). Succcssful orthogonal separations can be achieved when suitable mobile and stationary phases are selected, taking into account the physicochemical properties of the sample components including size and charge, hydrophobicity and polarity. Two typical examples of orthogonal combinations are the silver ion (SI) LC x non-aqueous (NA) RP-LC of triacylglycerols (separation according to degree of saturation and partition number) (20), and the NP-LC x RP-LC separation of essential oils, containing coumarins and psoralens (separation according to polarity and hydrophobicity (21).

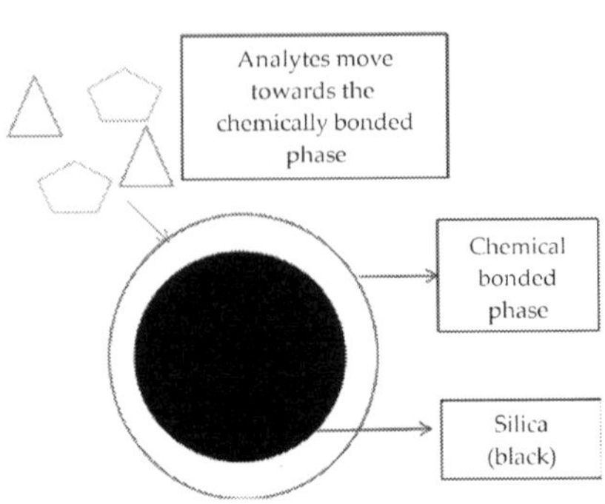

Figure 7.2: Principle of separation by HPLC

For the second dimension, the time for analysis (and regeneration when a gradient is applied) is determined by the flow rate used in the first dimension and the sampling rate. The second-dimension column internal diameter must be larger than the one of the column used for the primary separation (4-8 times larger (22). Fast second dimension must be performed at high flow rates within the sampling periods, ranging from 20 seconds to several minutes, depending on the application. Several approaches can be used to increase the analysis speed in HPLC.

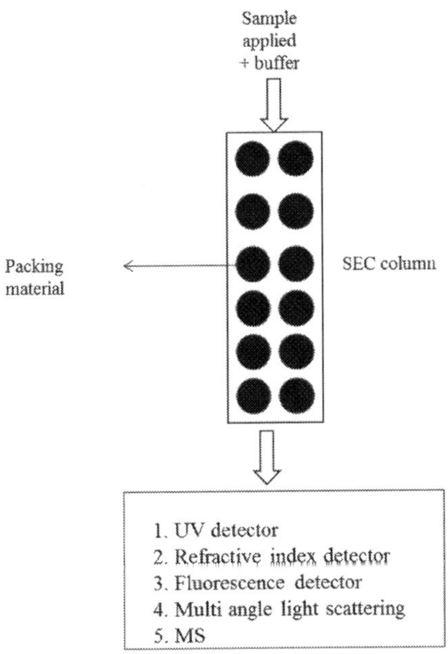

Figure 7.3: Mechanism of separation by HPLC

Monolithic columns

These columns have been extensively used in the second dimension, due to their short regeneration characteristics and high permeability (21-24). Furthermore, they allow operation at high flow rates without loss in resolution. Another way to speed up the second-dimension analysis is to use conventional columns packed with stationary phase of small particle size.

The use of sub-2μm particle packed columns at high flow rates in the second dimension leads to increased resolution, but it requires sophisticated instrumentation, which is capable of delivering high pressures. The flow rate can be strongly accelerated without loss in efficiency, through pressure drops, caused by the reduced viscosity of the mobile phase.

However, the stability of the stationary phase and analytes is of major concern, when elevated temperatures are used as a means to speed up the second-dimension analysis. Fused-core materials have been used for hyper-fast comprehensive separations with high resolution, for the analysis of polyphenolic antioxidants (25).

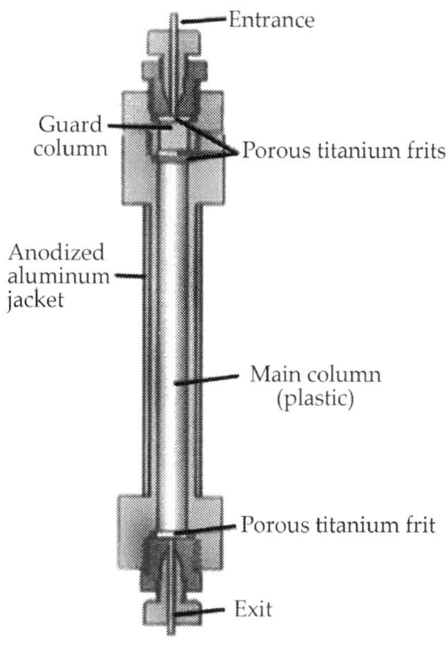

Figure 7.4: An HPLC column

Both isocratic and gradient programs are used in second dimension analyses. When ***isocratic elution*** is used, column conditioning, as well as the presence of negative effects due to the rapid changes of the modifier concentration on the background signals of UV or MS detectors are avoided. On the other hand, gradient elution in the second dimension is preferable, because it reduces the wrap-arounds. A wrap-around occurs when the retention time of a component in the second dimension exceeds the sampling time. This results in a broader peak, which affects the detection limit of the component.

Sometimes, several ***serially coupled columns*** can be used in the ***first dimension***, to reach high efficiency. In order to diminish signal interferences and reduce band broadening, the first dimension consists usually of micro- or narrow-bore column(s) to allow focusing. The low flow rates which are used with these columns provide fraction volumes that are compatible with the conventional bore columns of 4.6 mm i.e. in the second dimension. The interface ensures the automatic transfer of the primary column effluent to the second dimension in comprehensive LC systems. Several interface configurations have been designed and used in different applications. The loop interface is the most widely used interface and is based on a two-position /10-port (25,26), or 8-port (Hu, Chen, Kong et. al.,2005; Chen, Kong, Su et. al.,2004) switching valve, equipped with two storage loops with identical volumes. In this configuration, the loops are alternately emptied and filled with primary column effluent and emptied towards the second dimension in a continuous way.

In the ***packed loop interface*** the sampling loop can be modified by replacing the empty storage loops with loops packed with stationary phase. The mobile phase of the first dimension is preferably a weak solvent and the solutes are focused in the loop prior to their analysis in the second dimension. Fast desorption by a strong solvent should be performed by the secondary mobile phase. In the stop-flow interface approach, the columns are connected via an interface without sampling loops. One or several two-position/6-port switching valves have been used for this purpose. Focusing of the solutes at the head of the second-dimension column is necessary to avoid peak shape deterioration, therefore careful choice of the mobile phases in both dimensions is of paramount importance.

In the interface with ***parallel second dimension***, the second dimension operates with two columns to carry out analyses of consecutive fractions in parallel. Columns of the same batch and system configuration (adaptation of tubing lengths) should be used when a parallel second dimension is configured. In this approach, the additional use of elevated temperatures, under vacuum, enabled not only the collection of the fractions eluting from the first dimension via the interface valve, but also accomplished the evaporation of the mobile phase solvents.

MICRO-COLUMN AND NANOLIQUID CHROMATOGRAPHY:

Great efforts have been made in order to miniaturize LC instrumentation carrying out theoretical, technological and methodological studies. Better separation efficiencies, shorter analysis times due to the lower flow rates and successive coupling with MS have thus been achieved (29). Among these chromatographic techniques, ***microcolumn-*** and ***nanoLC*** have established themselves as complementary and/ or competitive separation techniques to conventional LC (30).

In 1967 Horvath and co-workers introduced ***microcolumn LC*** and used 0.5-1.0 mm I.D. stainless steel columns packed with pellicular particles for the separation of ribonucleotides. The minute sample sizes and the small volumetric flow- rates improve the detection performance when concentration –sensitive detectors are used, as a result of the considerably reduced chromatographic dilution (32). Nowadays, microcolumn LC is almost exclusively carried out as a research tool and in routine analysis, with slurry-packed columns of various dimensions (33). Solvent selection, slurry concentration, packing pressure, surfactants used to stabilize the particle suspension, column blank material and frit selection (34) are the most common parameters that influence the separation efficiency of microcolumns. Small I.D. packed columns possess an almost uniform crosssectional packing structure, therefore inhomogeneities in mobile phase flow paths are reduced (35). More rapid trans-column diffusion between all possible flow paths and retention regions is allowed, enabling the analytes to diffuse across the entire column cross-section (36).

Nano-LC was first introduced by Karlsson and Novothy in 1988 and since then it has been used as a complementary and/or competitive separation method to conventional HPLC, offering higher efficiency, shorter analysis time and better compatibility with MS, due to the relatively low flow-rates (40-600 nL min-1) that allow the transfer of the entire effluent from the column into the MS (37). Nano-LC is an environmentally safe and cost-effective technique. It employs minute volumes of mobile phases, producing minimum waste and also uses small amounts of stationary phases, which is beneficial when expensive packing materials are used. Nowadays, nano-LC, as a miniaturized version of HPLC brings significant improvements in analytical chemistry methods. It has been applied to the analysis of numerous compounds in different areas, such as pharmaceutical, environmental and proteomic, with the latter being the major application field, mainly because of the very low sample requirements. Nano-LC has been especially applied in protein sequencing, peptide mapping or determination of post-translational modification of proteins.

CHIP BASED LIQUID CHROMATOGRAPHY:

The demand of methods combining the possibility of analyzing minute samples in the shortest time possible and integrating in the very same instrument both sample pretreatment and separation has led to the development of micro-fabricated devices (named µ-total analysis system, µ-TAS). The main advantages of these devices are the low LODs, the increased peak capacity and sample loading and the minimization of the number of handling and transfer steps they offer. Many of these approaches, using centimeter-size glass, silicon or plastic chips have been successfully applied to proteomic analysis (38). Although the first µ-TAS chip was actually designed for pressure driven LC (39), most of the on-chip separations developed up to now have been focused on electrokinetically driven separation techniques such as Capillary Electrophoresis (CE), Capillary Electrochromatography (CEC) and Micellar Electrokinetic Chromatography (MECK).

A number of problems that have hindered the development of HPLC chips are among others the following: the difficulty of integrating on-chip injectors and mechanical valves, the lack of easy flow control, the on-chip application of high pressures and the manufacturing of columns with sufficient sample capacity (40). However, the situation with chip-based LC has been changing and several approaches have been developed which show that the technological problems, at least partly, have been solved. For instance, the system is commercially available. Both systems make use of columns packed with commercially available beads. The pressure is applied by an external pump in Yin's device and by an integrated pump based on electrolysis in the system. Although these systems offer the advantages related to a chip-based format, i.e., low dead-volume on-chip connections, suitability for small volume samples and ease of automation, they still miss an important piece of progress that can be made by micromachining, namely an entirely micromachined separation column.

Micromachining is needed to optimize column performance. Most columns operate at plate heights around hmin (where h, plate height) (41). Much lower values of hmin could be obtained if the stationary phase were made entirely ordered, for instance by micromachining the entire column. The open tubular column, which would be the ultimate ordered column, unfortunately does not offer sufficient sample capacity to be practically useful. Several technological challenges remain before micromachined columns can become standard components in chip-based LC systems. Thus, partly porous columns must be manufactured in order to ensure sufficient sample capacity. It is also very important to have a uniform coating of the stationary phase, because non-uniformities in the coating will almost have the same detrimental effect on the plate height as nonuniformities in the flow-path. One way of homogenizing the flow-path, further than allowed by particle packing, is by using organic or inorganic monoliths as the stationary phase. Monolith development for chip-based LC is therefore an active research area (42).

Another technological challenge is the application of high pressures to the chip, a problem which is almost solved, taking into account the interfacing with external pumps. A further hindrance is the necessity to inject a narrow plug of analyte onto the column, in order to limit the contribution of the injection to plate height. Several interesting approaches have been reported, as for instance, that of Reichmuth et al., employing a hydrophobic valve and those of the off-line injection valves and the on-column injection channel or slit.

ULTRA-PERFORMANCE LIQUID CHROMATOGRAPHY:

Ultra-Performance Liquid Chromatography (UPLC) can be considered to be a relatively new direction of liquid chromatography. Throughout the history of HPLC there has been a trend to use smaller particles as packing material. It is well known that as particle size decreases to less than 2.5 μm, there is a significant gain in efficiency, and thus resolution, and the efficiency does not diminish at increased linear velocities or flow rates according to the Van Deemter equation. By using smaller particles, speed and peak capacity (number of peaks resolved per unit time) can be extended to new limits which is known as Ultra Performance. UPLC takes advantage of the huge progress achieved in particle chemistry performance, system optimization, detector design and data processing and control. The use of sub-2 μm particles and mobile phases at high linear velocities and instrumentation that operates at higher pressures than those used in HPLC, resulted in dramatic increases in resolution, sensitivity and speed of analysis.

This new category of analytical separation science retains the practicality and principles of HPLC while creating a step function improvement in chromatographic performance (Swartz, 2004). In conventional HPLC the use of continuously smaller sized packing material particles leads to higher column back-pressure. Small column diameters like 2.1 or 1.0 mm could also cause similar problems and disable their use under the conventional conditions. In order to overcome the pressure limitation of conventional equipment, shorter columns packed with small particle diameter particles were used. However, the use of UPLC on a routine basis in the laboratory, required further improvement in some practical aspects, such as sample introduction, reproducibility and detection. Ultra high Ultra performance Liquid Chromatography (UPLC) pressure columns required extremely narrow sample plugs to minimize any sample volume contribution to peak broadening. The advent of UPLC demanded also the development of a new instrumental system for liquid chromatography, which would take advantage of the separation performance (by reducing dead volumes) and withstand the pressures (about 8,000 to 15,000 psi, compared with 2,500 to 5,000 psi in HPLC). Efficiency is proportional to column length and inversely proportional to the particle size. Therefore, the column can be shortened by the same factor as the particle size without loss of resolution.

OPERATION

Mobile phase enters the column from the left, passes through the particle bed, and exits at the right. Flow direction is represented by green arrows. First, consider the top image; it represents the column at time zero [the moment of injection], when the sample enters the column and begins to form a band. The sample shown here, a mixture of yellow, red, and blue dyes, appears at the inlet of the column as a single black band (In reality, this sample could be anything that can be dissolved in a solvent; typically, the compounds would be colorless and the column wall opaque, so we would need a detector to see the separated compounds as they elute).

After a few minutes (lower image), during which mobile phase flows continuously and steadily past the packing material particles, we can see that the individual dyes have moved in separate bands at different speeds. This is because there is a competition between the mobile phase and the stationary phase for attracting each of the dyes or analytes. Notice that the yellow dye band moves the fastest and is about to exit the column. The yellow dye likes (is attracted to) the mobile phase more than the other dyes. Therefore, it moves at a *faster* speed, closer to that of the mobile phase. The blue dye band likes the packing material more than the mobile phase. Its stronger attraction to the particles causes it to move significantly *slower.* In other words, it is the most retained compound in this sample mixture. The red dye band has an intermediate attraction for the mobile phase and therefore moves at an *intermediate* speed through the column. Since each dye band moves at different speed, we are able to separate it chromatographically. This technique is discussed in details in chapter 8.

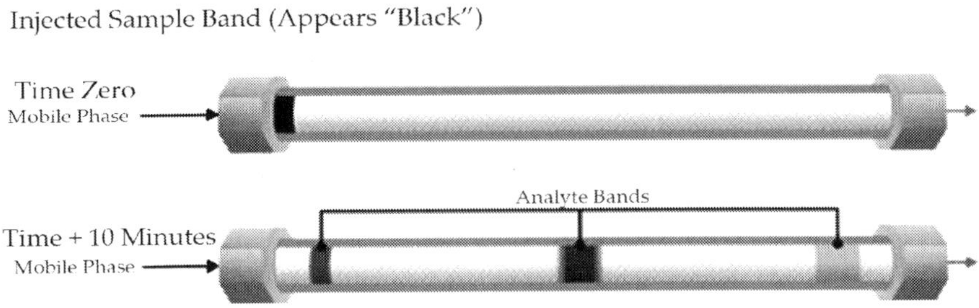

Figure 7.5: Understanding How a Chromatographic Column Works – Bands

TYPES OF HPLC

Types of HPLC generally depend on phase system used in the process. Following types of HPLC generally used in analysis

Normal Phase Chromatography:

Also known as Normal Phase HPLC (NP-HPLC), this method separates analytes based on polarity. NP-HPLC uses a polar stationary phase and a non-polar mobile phase. The polar analyte interacted with and is retained by the polar stationary phase. Adsorption strengths increase with increased analyte polarity, and the interaction between the polar analyte and the polar stationary phase increases the elution time.

Reverse Phase Chromatography:

Reversed phase HPLC (RP-HPLC or RPC) has a non-polar stationary phase and an aqueous, moderately polar mobile phase. RPC operates on the principle of hydrophobic interaction, which result from repulsive forces between a polar eluent, the relatively non-polar analyte, and the non-polar stationary phase. The binding of the analyte to the stationary phase is proportional to the contact surface area around the non-polar segment of the analyte molecule upon association with the ligand in the aqueous eluent.

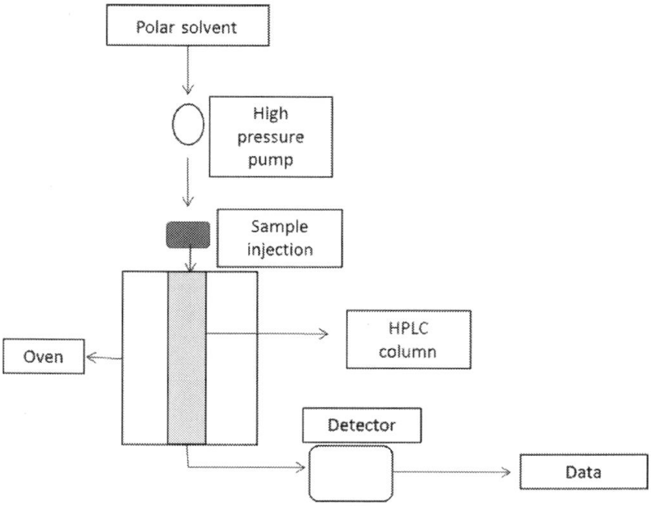

Figure 7.6: Schematics of RP-HPLC

Size Exclusion Chromatography:

Size exclusion chromatography (SEC), also called as gel permeation chromatography or gel filtration chromatography mainly separates particles on the basis of size. It is also useful for determining the tertiary structure and quaternary structure of proteins and amino acids. This technique is widely used for the molecular weight determination of polysaccharides.

Ion-Exchange Chromatography:

In ion-exchange chromatography, retention is based on the attraction between solute ions and charged sites bound to the stationary phase. Ions of the same charge are excluded. This form of chromatography is widely used in purifying water and proteins. High-pH anion-exchange chromatography is used for purification of carbohydrates and oligosaccharides (44,45).

Bio-affinity Chromatography:

Separation based on specific reversible interaction of protein with ligands. Ligandsare covalently attached to solid support on a bio-affinity matrix, retains proteins with interactions to the column-bound ligands.

Proteins bound to a bioaffinity column can be eluted in two ways:

- Biospecific elution: Inclusion of free ligand in elution buffer which competes with column bound signal.
- Aspecific elution: Change in pH, salt etc. which weakens interactions protein with column-bound substrate.

Because of specificity of the interactions, bioaffinity chromatography can result in very high purified in a single step (10-1000 fold).

PARAMETERS

For the accurate analysis of a compound, there are some parameters which are used as a standard for a particular compound. If there is a change occurs in the parameters the result may be affected greatly. The most commonly used parameters are internal diameter, particle size, pore size, pump pressure. For different compounds, the parameters can be changed according to their nature and chemical properties.

Internal diameter:

The internal diameter (ID) of an HPLC column is a critical aspect that determines quantity of analyte that can be loaded onto the column and also influences sensitivity. Larger columns are usually seen in industrial applications such as the purification of a drug product for later use. Low ID columns have improved sensitivity and lower solvent consumption at the expense of loading capacity.

Particle size:

Most traditional HPLC is performed with the stationary phase attached to the outside of small spherical silica particles (very small beads). Smaller particles generally provide more surface area and better separations, but the pressure required for optimum linear velocity increases by the inverse of the particle diameter squared.

Pore size:

Many stationary phases are porous to provide greater surface area. Small pores provide greater surface area while larger pore size has better kinetics especially for larger analytes. Pore size defines an ability of the analyte molecules to penetrate inside the particle and interact with its inner surface. This is especially important because the ratio of the outer particle surface to its inner one is about 1:1000. The surface molecular interaction mainly occurs on the inner particle surface.

Pump pressure:

Pumps vary in pressure capacity, but their performance is measured on their ability to yield a consistent and reproducible flow rate. Modern HPLC systems have been improved to work at much higher pressures, and therefore be able to use much smaller particle sizes in the columns (< 2 micrometres).

APPLICATIONS OF HPLC

The information that can be obtained using HPLC includes identification, quantification, and resolution of a compound. Preparative HPLC refers to the process of isolation and purification of compounds. This differs from analytical HPLC, where the focus is to obtain information about the sample compound.

Chemical Separations:

It is based on the fact that certain compounds have different migration rates given a particular column and mobile phase, the extent or degree of separation is mostly determined by the choice of stationary phase and mobile phase.

Purification:

Purification is defined as the process of separating or extracting the target compound from a mixture of compounds or contaminants. Each compound showed a characteristic peak under certain chromatographic conditions. The migration of the compounds and contaminants through the column need to differ enough so that the pure desired compound can be collected or extracted without incurring any other undesired compound.

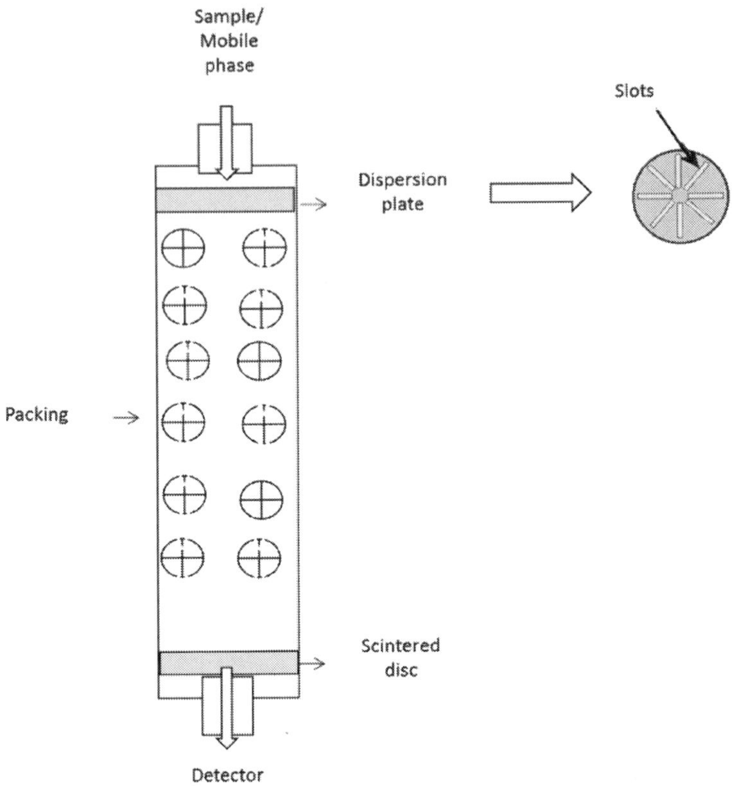

Figure 7.7: A preparative chromatography set up

Identification:

Generally, assay of compounds is carried using HPLC. The parameters of this assay should be such that a clean peak of the known sample is observed from the chromatograph. The identifying peak should have a reasonable retention time and should be well separated from extraneous peaks at the detection levels which the assay will be performed.

Other applications of HPLC include:

Pharmaceutical applications (45)

- Tablet dissolution study of armaceutical dosages form.
- Shelf-life determinations of parmaceutical products

- Identification of active ingredients of dosage forms
- Pharmaceutical quality control

Environmental applications (46)

- Detection of phenolic compounds in Drinking Water
- Identification of diphenhydramine in sedimented samples
- Bio-monitoring of pollutant

Forensics (47)

- Quantification of the drug in biological samples.
- Identification of anabolic steroids in serum, urine, sweat, and hair
- Forensic analysis of textile dyes.
- Determination of cocaine and metabolites in blood

Clinical (48)

- Quantification of ions in human urine Analysis of antibiotics in blood plasma.
- Estimation of bilirubin and bilivirdin in blood plasma in case of hepatic disorders.
- Detection of endogenous neuropeptides in extracellular fluids of brain.

Food and Flavor (49)

- Ensuring the quality of soft drink and drinking water.
- Analysis of beer.
- Sugar analysis in fruit juices.
- Analysis of polycyclic compounds in vegetables.
- Trace analysis of military high explosives in agricultural crops.

LIMITATIONS OF HPLC

- Lack of efficiency compared with other techniques.

- Low diffusion coefficients in the liquid phase involved in slow diffusion of analytes in the stationary phase results in long analysis time. Therefore, the evaluation of the HPLC technique has led to the reduction in the size of the packing materials.
- As the size of the packing decreased, the efficiency, resolution and the ability to work with short analysis time increased, resulting in use of different dimension of column.

CONCLUSION

The HPLC is mostly used analytical technique. It is having several advantages. With the use of HPLC one can produce extremely pure compounds. It can be used in both laboratory and clinical science. With the use of HPLC the accuracy, precision and specificity can be increased. The only disadvantage of HPLC is high cost.

REFERENCES

1. Rogatsky E. Modern high performance liquid chromatography and HPLC 2016 International Symposium. J Chromatogr Sep Tech. 2016;7:e135.
2. Mulubwa M, Rheeders M, Du PL, Grobler A, Viljoen M. Development and validation of high performance liquid chromatography tandem mass spectrometry (HPLC-MS/MS) method for determination of tenofovir in small volumes of human plasma. J Chromatogr Sep Tech. 2015;6:300.
3. Santini DA, Sutherland CA, Nicolau DP. Development of a high performance liquid chromatography method for the determination of tedizolid in human plasma, human serum, saline and mouse plasma. J Chromatogr Sep Tech. 2015;6:270.
4. AL-Jammal MKH, Al Ayoub Y, Assi KH. Development and validation of micro emulsion high performance liquid chromatography (MELC) method for the determination of nifedipine in pharmaceutical preparation. Pharm Anal Acta. 2015;6:347.
5. Shephard GS, Sydenham EW, Thiel PG, Gelderblom WC . Quantitative determination of fumonisins b1and b2 by high-performance liquid chromatography with fluorescence detection. J

Liquid Chromatogr. 2006:13.

6. Hamscher G, Sczesny S, Höper H, Nau H.. Determination of persistent tetracycline residues in soil fertilized with liquid manure by high performance liquid chromatography with electrospray ionization tandem mass spectrometry. Anal Chem.2002;74:1509-1518.
7. Svec F, Frechet MJJ. Continuous rods of macroporous polymer as high-performance liquid chromatography separation media. Anal Chem. 1992;64:820-822.
8. Shintani H. Validation Study in membrane chromatography adsorber and phenyl hydrophobic membrane chromatography adsorber for virus clearance and removal of many other components. PharmAnalActa. 2013;S2:005.
9. Ramsteiner KA. Systematic approach to column switching. J Chromatogr. 1988;456, 3.
10. Little CJ, Tompkins DJ, Stahel O. Contemporary Practice of Chromatography. J Chromatogr. 1983;264, 183.
11. Little CJ, Stahel O. Increased sample throughput in HPLC using sample switching. Chromatographia. 1984;19, 322
12. Patrick DW, Kracht WR. Single-valve single-pump high performance liquid chromatography column-switching analysis of hydrolyzed wood components. J Chromatogr.1985;318, 269 (1985)
13. Hsu S-H. Zone compressesion effects in high performance liquid chromatography. J. Chromatogr A 1986;367, 293.
14. Dugo P, Cacciola F, Kumm T, et. Al. Comprehensive multidimensional liquid chromatography: theory and applications. J Chromatogr A. 2008;1184, 353.
15. Guiochon G, Marchetti N, Mriziq K, et. al. Implementation of two-dimensional liquid chromatography. J Chromatogr A. 2008;1189, 109.
16. Francois I, Sandra K, Sandra P. Comprehensive liquid chromatography: fundamental aspects and practical considerations-a review. Anal Chim Acta. 2009;641, 14.
17. Davis JM, Giddings JC. Statistical theory of component overlap in multicomponent chromatograms. Anal Chem. 1983;55, 418.

18. Giddings JC. Sample dimensionality: A predictor of order-disorder in component peak distribution in multidimensional separation. J Chromatog A. 1995;703, 3.
19. Liu Z, Patterson Jr DG, Lee ML. Geometric approach to factor analysis for the estimation of orthogonality and practical peak capacity in comprehensive two-dimensional separations. Anal Chem. 1995;67, 3840.
20. Mondello L, Tranchida, Stanek et al. Silver-ion reversed-phase comprehensive two-dimensional liquid chromatography combined with mass spectrometric detection in lipidic food analysis. J Chromatogr A.2005;1086, 91.
21. Schoenmakers PJ, Vivó-Truyols G, Decrop WMC. Advances in chromatography. J Chromatogr A. 2006;1120, 282.
22. van Nederkassel AM, Aerts A, Dierick A, et. al. Fast separation on monolithic silica columns: method transfer, robustness and column ageing for some case studies. J Pharm Biomed Anal. 2003;32, 233.
23. van Nederkassel AM, Aerts A, Dierick A, et. al. Performance limits of monolithic and packed capillary columns in high-performance liquid chromatography and capillary electrochromatography. J Pharm Biomed Anal. 2003;32, 233.
24. Dugo P, Herrero, Kumm et. al. Comprehensive normal *reversed-phase liquid chromatography coupled to photodiode array and mass spectrometry detection for the analysis of free carotenoids and carotenoid esters from mandarin. J Chromatogr A. 2008;1189, 196.
25. Dugo P, Cacciola F, Herrero M, et. Al. Use of partially process column as second dimension in comprehensive two-dimensional system for analysis of polyphenolic antioxidants. J Sep Sci. 2008;31, 3297.
26. Jandera P, Lahovska F, et al. Two-dimensional liquid chromatography normal-phase and reversed phase separation of (co)oligomers. J Chromatography A. 2006;1119, 3.
27. Hu L, Kong C, et al. Improved performance of comprehensive two-dimensional HPLC separation of traditional Chinese medicines by using a silica monolithic column and normalization of peak heights. J Chromatogr A. 2005;1092, 191.
28. Chen X, Su K, et al. Separation and identification of compounds

in Rhizoma chaunxiong by comprehensive two-dimensional liquid chromatography coupled to mass spectrometry. J Chromatogr A. 2004;1040, 169.

29. Hernández-Borges J, Aturki Z, Rocco A, et. al. Recent applications in nanoliquid chromatography. J Sep Sci. 2007;30, 1589.
30. Karlsson KE, Novotny M. Separation efficiency of slurry-packed liquid chromatography microcolumns with very small inner diameters. Anal Chem. 1988;60, 1662.
31. Horvath CG, Preiss BA, Lipsky SR. Fast liquid chromatography: an investigation of operating parameters and the separation of nucleotides in pellicular ion exchangers. Anal Chem. 1967;39, 1422.
32. Novotny M. Recent advances in microcolumn liquid chromatography. Anal Chem. 1988;60, 500A.
33. Meyer RF, Hartwick RA. Efficient packing of small particle microbore columns. Anal Chem. 1984;56, 2211.
34. Zimina TM, Smith RM, Myers P, et. al. Effects of kinematic viscosity of the slurry on the packing efficiency of PEEK microbore columns for liquid chromatography. Chromatographia. 1995;40, 662.
35. Kennedy RT, Jorgenson JW. Preparation and evaluation of packed capillary liquid chromatography columns with inner diameters 20 to 50 µm. Anal. Chem. 1989;61, 1128.
36. Hsieh S, Jorgenson JW, Preparation and evaluation of slurry-packed liquid chromatography microcolumn with inner diameters from 12 to 33µm. Anal Chem. 1996;68, 1212.
37. Tomer KB. Separations combined with mass spectrometry. Chem Rev. 2001;101, 297.
38. Ghitun M, Fortier B, et al. Integrated microfluidic devices with enhanced separation performance: Application to phosphoproteome analyses of differentiated cell model systems. J Sep Sci. 2006;29, 1539.
39. Manz A, Miura M, et al. Design of an open-tubular column liquid chromatograph using silicon chip technology. Sens Actuators B1. 1990;249.
40. Eijkel J. Chip-based HPLC: the quest for the perfect column. Lab Chip. 2007;7, 815.

41. Knox JH. Band dispersion in chromatography- a new view of A-term dispersion. J Chromatogr A. 1999;831, 3.
42. Reichmuth DS, Shepodd TJ, Kirby BJ. Microchip HPLC of peptides and proteins. Anal Chem. 2005;77, 2997.
43. Swartz M.E. UPLC:An introduction and review. J Liq Chromatog. & Related Technol. 2005;28, 1253.
44. Abidi SL. High-performance liquid chromatography of phosphatidic acids and related polar lipids. J Chromatogr. 1991;587: 193-203.
45. Hearn MTW. Ion-pair chromatography on normal and reversed-phase systems. Adv Chromatogr. 1980; 18: 59–100.
46. Lauback RG, Rice JJ, Bleiberg B. et. al. Specific High-performance Liquidchromato- graphic Determination of Ampicillin in Bulks, Injectables, Capsules and Oral Suspensions by Reversed-phase Ion-pair Chromatography. J Liq Chromatogr. 1984;7(6): 1243-1265.
47. Haginaka J, Yasuda H, Uno T, et. al. Alkaline degradation and determination by high-performance by high-performance liquid chromatography. Chem Pharm Bull. 1984;32: 2752-2758.
48. Abidi SL. High-performance liquid chromatography of phosphatidic acids and related polar lipids. J.Chromatogr. 1991; 587: 193-203
49. Christie WW, Gill S, Nordbäck J, Itabashi Y, Sanda S, Slabas AR. New procedures for rapid screening of leaf lipid components from *Arabidopsis*. Phytochemical Anal. 1998; 9: 53-57

CHAPTER 8

ULTRA PERFORMANCE LIQUID CHROMATOGRAPHY

INTRODUCTION

The term UPLC refers to Ultra Performance Liquid Chromatography that offers several advantages related to: the resolving power, rapidity and its higher sensitive detection. The technique employs the use of granulated particles of fine dimensions and causes lower use of solvent that is applied on the column.

The technique of UPLC is derived from High Performance Liquid Chromatography (HPLC): with alterations in the packing materials. The methodology has been applied for three decades; across several institutes in the world. A tenet of HPLC dictates that a reduction in size of the particle size of the columns to cause an enhancement in the packing efficiency and resolving power (1).

WHY UPLC?

The protocol of HPLC presents advantages that include an increased sensitive and selective detection as well as its sturdiness but is challenged by a decreased efficiency when looked at with GC (2).

UPLC uses an operational pressure more than that of HPLC while the size of particles is also less than 2.5 µm. On the basis of Van Deemter equation, a reduction in size of particles below 2 micro meters causes an increase in efficiency with a boost in flow rates and velocities not affecting the performance (3).

The peaks reported as a function of time can be maximized by employing particles that are less in size along with an increase in speed in the system as compared to HPLC. A further increase in the performance of the system can be accomplished by the use of:

A boost in temperature that is effects an enhancement in the flow rate of the mobile phase that in turn causes a dip in its viscosity that further limits the challenges of back pressure.

The use of a solid piece with a slew of paths to flow via pores in its skeleton: such an apparatus is called Monolithic columns. These materials found application in gels that were applied for polymers that are organic in nature. The columns are composed of packed polymers that offer a decrease in friction to flow as compared to traditional ones.

A combination of both the above principles aids in the design of UPLC that has the advantages of:

- Increased sensitivity
- Rapid time
- Enhanced speed
- A boost in the resolution
- A dip in the quantity of solvent required
- When compared to HPLC systems (1).

PRINCIPLE OF OPERATION

The technique uses particles of dimensions that are below 2.5 µm as compared to a particle size of three to five microns in High Performance Liquid Chromatography. The principle is based on the Van Deemter equation (3):

$$H = A + \frac{B}{v} + Cv$$

The terms, A, B and C are constants:

A does not depend on velocity is attributed to eddy effects with lowest values reported with the use of small particles of uniform nature.

B is attributed to axial diffusion that is connected with the diffusion potential of samples. As it decreases with an increase in flow rates, a term for velocity: v is divided from it.

C is attributed to kinetic resistance to equilibrium during the separation of samples. It represents a pause in time when samples shuttle between the stationary and mobile phases. The value of C increases as flow of a gas increases: expressed as directly proportional with the velocity.

Looking at the equation, an increase in the analysis output can be effected with minimum compromise on the efficiency. The use of novel apparatus and systems in order to modify the available instrumentation to increase pressures involved and a decrease in the dead volumes due to smaller particle size.

It has also been reported that the efficiency varies inversely with the size of particles and directly with the length of the column (4).

Apart from the factor of small particles, the system warrants the use of novel additions in the apparatus used. It started off with descriptions of the use of pumps to boost the pressures involved that led to the development of such systems (3).

ACQUITY UPLC™ was one of the first apparatus to be designed on a commercial scale to be applied on proteins and other pharmaceutical compounds. It possessed a binary solvent manager that combines the output of 2 pumps to pump in a mixture at elevated pressures.

PARTICLES USED

The biotechnology industry has synthesized particles that were of 1.5 μm dimensions that were not porous. However, these particles could not be applied on UPLC as the higher pressures necessitate the use of porous particles. Additionally, these particles could not accommodate much samples as their area was lesser. This led to the development of silica that was strong but presented challenges of the formation of tails in the chromatogram. The technique of chemistry led to the development of columns that were found to be efficient by the use of methyl additives. Though the columns lacked stability, they were strong and efficient (1).

This led to the development of columns that were based on bridged ethane hybrid (BEH) technology. The technique uses a link of two methyl moieties embedded in silica to develop columns that were strong as resistant to pressure. Due to their increased packing in column to increase length and small particle size, the technology has found application in the analysis of several molecules with enhanced resolution (5).

INSTRUMENTATION

The system is composed of several features that include:

Instruments that can withstand elevated pressures

Particles of little dimensions that can withstand elevated pressures

Detectors

The use of software that is integrated

The volume of the system is very low.

For instance, ACQUITY UPLC™ systems have been reported to offer easy analysis and operation along with simple operations, heightened production and data that can be used on software such as MassLynx™ and Empower™ (Acquity UPLC column).

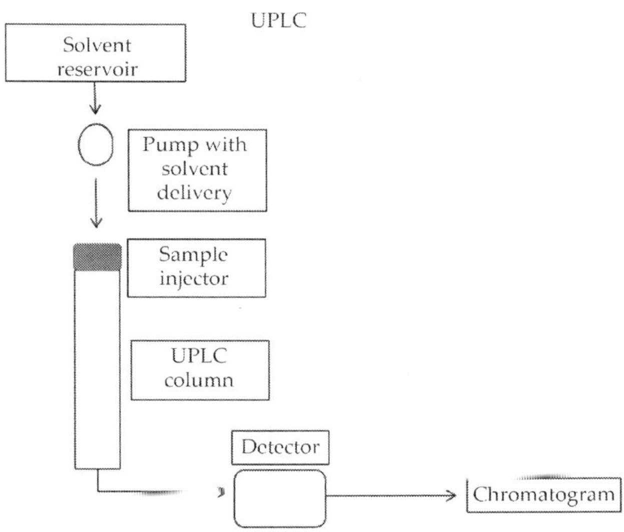

Figure 8.1: UPLC Instrumentation

PUMPS

A pump to be used in the system should be capable of reaching an applying pressure of 15000 psi to achieve a satisfactory flow rate for a column that is packed with particles of dimensions 1.7 μm to accommodate for the pressure drop across a column that is 15 centimeters. Pumps can also be operated in tandem each of a ceiling value of 1000 bar along with the use of valves that can select the appropriate amounts of solvents to be applied from four different inlets (1)

SAMPLE INJECTORS

The design of appropriate sample injectors has been reported in literature (6,7). The introduction of samples is a key feature as the particles used in the column are of micro ranges. The use of a sample manager based on a needle design achieves three main ends:

- The injection of samples without pulses to avoid damage to the column
- A sturdy fastening of needle
- A decrease in any form of broadening of bands

In order to achieve the process of sample injection, the following steps are reported that include:

A valve called inject valve collects samples from its container, the requisite volume of sample is withdrawn from the vial following which the needle enters the injection port. This is followed by an "on mode" of injection valve caused by pressing of the needle and the injection port to release the sample. A thorough wash for optimum periods aids in the reuse of the injector without contamination.

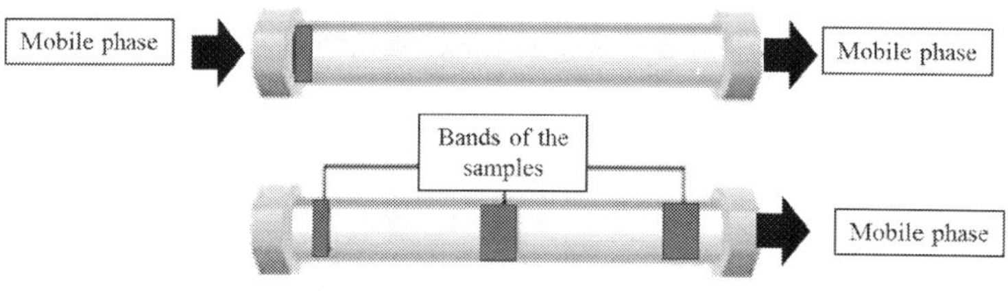

Figure 8.2: Separation using UPLC columns

UPLC COLUMNS

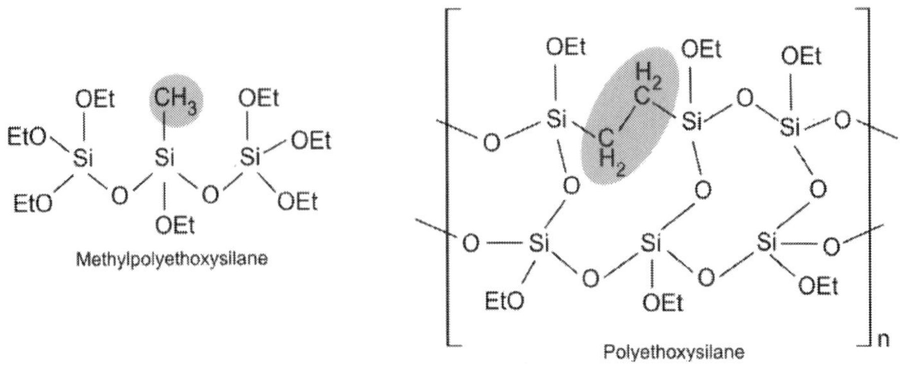

Figure 8.3: Bridged Ethyl Hybrid (BEH) technology.

From the use of non-porous molecules to the use of silica, the year 2000 saw the development of XTerra® that combined polymers with silica. The material has high strength along with effiecincy over several conditions yet lacked stability that led to the development of Bridged Ethyl Hybrid (BEH) technology.

These particles are Packing 1.7 μm and are assembled into a column with smooth surfaces in a uniform manner. The particles are embedded within a matrix to serve as a stationary phase. There are 4 varieties of stationary phases for UPLC synthesized by ACQUITY:

- BEH C18 and C8 columns: They are the most in vogue and are functional over a vast pH. They are composed of straight chains of alkyl groups.
- BEH Shield R18 columns: A feature of increased selection is presented with complementation to the above.
- BEH Phenyl columns: The presence of tri-functional C6 alkyl ethyl between the phenyl rings and the silyl moieties is the hallmark of the column.
- BEH Amide columns: These columns have extended stabilities to be stored for a long time and are operational over a wide ph from 2 up to 12. They are composed of BEH particles with trifunctionally bonded amides.

The presence of eCord™ microchip aids in the quality heck of each column in instruments such as ACQUITY UPLC BEH. They offer advantages such as Increased stability to pH, more efficiency and a boost in the strength

COLUMN MANAGER

In a system of ACQUITY UPLC, the presence of a column manager is a hallmark of the system. The system maintains the temperature within the column 282-393K along with a facility for shifting between several columns (up to four).

There is a solvent manager that allows the use of two pumps to supply a gradient that is adjusted according to requirements. The sample manager enables the injection of samples using pressure to avoid sample dispersion apart from the use of sensors to maintain the levels in an accurate fashion. A 25s time for injection followed by a minute entry and washing maintains purity of samples.

DETECTORS

The detectors in this system face a challenge as the use of particles with small dimensions yields peaks whose dimensions are below a second. The detectors used should have an enhanced sensitivity as compared to HPLC ones apart from the use of very less dispersion.

Two UV-visible photodiode array detectors that have been reported in literature include: ACQUITY PDA and ACQUITY PDA eλ. These detectors have reduced levels of noise (10 Au) while the rate of data is 80 Hz . The use of Teflon AF flow cells causes a boost in the efficiency by circumventing the phenomenon of internal reflection.

ACQUITY Tunable UV/Visible detector is another feature in the system that aids in the efficient transmission of light along with the use of detectors that avoid leaking and dispersion (1,6).

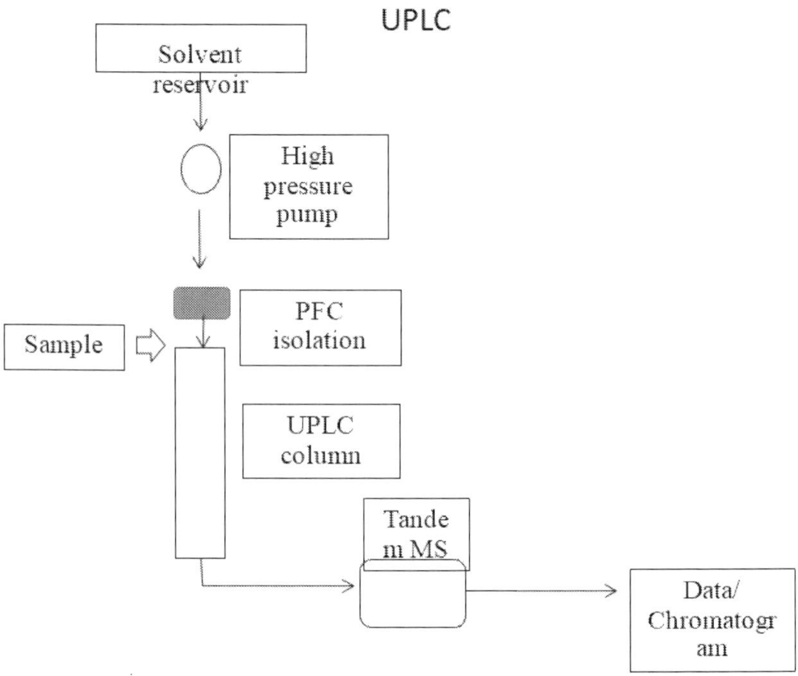

Figure 8.4: Scheme of UPLC

The amalgamation of UPLC with tandem mass spectrometry: (UPLC™-MS/MS) has been reported to possess an enhanced resolving power, sensitivity and rapidity

For instance, a 2006 study showed that the system could efficiently detect pesticides in ground water samples with high precision (8).

UPLC and HPLC

Both the techniques are based on liquid chromatography, with HPLC that applies enhanced pressure on the mobile phase through the column, used extensively for the purpose of analysis, purification and quantification of mixtures. The technique of HPLC offers advantages that include

UPLC vs HPLC

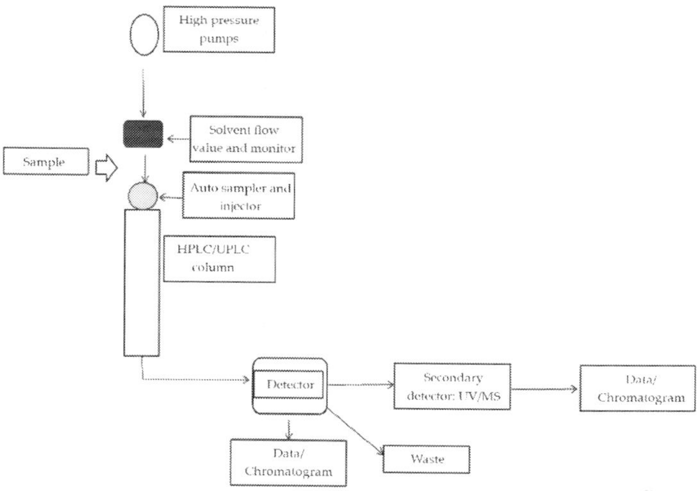

Figure 8.5: UPLC versus HPLC

- An enhanced resolving power
- The reuse of columns is permitted Rapid analysis
- Can be subjected to easier scale up

The technique of UPLC is an offshoot of HPLC with particles used with dimensions below 2um. The system was patented by Waters Corporation yet its use has been tested by several labs across the world.

A main difference in the techniques is the use of higher pressures: HPLC uses pressures of 170 to 350 bars while UPLC employs pressures of 500 to 1000 bars.

UPLC ADVANTAGES

There are reports of a reduction in the time of the analysis using UPLC columns as compared to other techniques (9).

- The amount of solvent consumed is lesser due to the decrease in the time of processing.
- The technique has an enhanced resolution reported. For instance, a run of the drug Metoprolol on an HPLC column

reported a resolution of 2.7 while the value was 26.9 for UPLC (1).

- There is an increase in the output of samples that encourages the industry to apply this technique to scale up operations and avoid batch issues.
- The technique offers the use of multi residue methods: an application to waste water is discussed in the applications section.
- The technique can be applied to quantification of compounds.
- An analysis in real time is also presented by the technique.
- It offers products that have assured quality (1).

Disadvantages of UPLC

- The technique employs elevated pressures of more than 1000 bars that adversely affect the performance of a column in addition to adding to its maintenance (10).
- The particles of dimensions 1.7 μm are reportedly non-reusable (1)

The following table lists the points of differences between various parameters between the two techniques:

Table 1: Comparison Between UPLC and HPLC

Characteristics	**UPLC**	**HPLC**
Column dimensions	50 X 2.1 mm	150 X 3.2 mm
Total run time	1.5 min	10 min.
Injection volume of sample: Less sample required	2μL	5μL
Pressure: More pressure: Rapid separation	1000 bars	300-400 bars
Size of packing particles: A dip in time for analysis	Less than 2μm	3 to 5μm

Numbers of plates generated	7500	2000
Column temperature: Decreases solvent viscosity and boosts the transfer rate of mass	65°C	30°C
Flow rate of mobile phase	Greater to minimize the level used.	Lesser

APPLICATIONS OF UPLC

Identification of metabolites:

The establishment of identity of a new chemical entity (NCE) warrants its study for further use. For instance, a 2005 study examined the use of UPLC-MS for drugs such as acetaminophen. The technique was reported as sensitive with high resolution and specificity (11). Several drugs have been tested on a UPLC column that can be summarized as below:

Table 2: Drugs tested on UPLC

Drug tested	Retention time (min)
Secnidazole	3.7
Fluconazole	4.1
Azithromycin	6.2
Diclofenac	1.5
Aspirin	0.83

Identification of Metabolites

Due to the higher sensitivity and accuracy, UPLC-MS/MS has been reported as very useful in the field of biomarker discovery:

- A 2013 study examined metabolites associated with gentiopicroside using UPLC along with UPLC/Q-TOF MS and NMR (12).
- A 2015 study examined various metabolites produced as a result of administering a species of yam used as Chinese medicine: Dioscorea saponins (13).

- The detection of nicotine with several of its metabolites was performed using UPLC with MS for further kinetic studies (14).

Detection of impurities:

HPLC had been reported to possess high sensitivity towards the detection of trace amounts of impurities that were reproducible. However, the necessity to repeat the analysis is challenging due to the use of excipients. The presence of an impurity with a similar stricture and mass as an analyte cannot be detected by techniques such as LC-MS however showed good detection on UPLC (15).

The drug development and formulae demand accurate measurement of low-degree impurities in the energetic pharmaceutical ingredients or the excipients or the uncooked substances used in the the final product. Thus, the combination of UPLC with mass spectrometry has been useful for the documentation of drug and endogenous metabolites within the final product.

A 2015 study reported the application of UPLC for the characterization of impurities present in halobetasol propionate cream (16).

Dissolution testing:

The quality check of a drug before release requires its testing for dissolution. This step serves as both a rate limiting step as well as indicative of batch uniformity. The dissolution profile of a drug signifies reliability and batch to batch uniformity of API (Active Pharmaceutical Ingredients) inside the formulations. Nowadays, novel formulations are coming within the marketplace which calls for stronger analytical sensitivity. This can be fulfilled by way of UPLC which gives on line pattern acquisition.

- A 2011 study examined the use of UPLC for the dissolution testing of a preparation of rabeprazole sodium (17).
- By the usage of UPLC, dissolution trying out may be completely automatic as it could carry out features like records acquisition, evaluation of samples aliquots, control of take a look at effects and distribution (1).

Bioequivalence studies:

The use of LC MS has been documented for detection of drugs in materials. The increased sensitivity and resolution supports the use of UPLC in the analysis of bioequivalence. Bioequivalence studies are pharmacokinetic research needed for the quantitation of medicine in organic samples. This is an essential step to examine the charge and publicity level of newly evolved formulations of triumphing drugs with that of the authentic components. The selectivity and sensitivity of UPLC-MS/MS produce reliable and particular information. UPLC-MS/MS answers have extended efficacy, output, and profitability for the bioequivalence laboratories. UPLC pattern manager complements the effectiveness by way of considering a big variety of samples in a temperature controlled system, confirming most throughput which will increase the sensitivity and excellent of records acquisition of tandem quadrupole MS systems.

- A study in 2001 reported that UPLC showed a 20-fold enhancement in results as compared to nanoelectrospray. Using protein samples, the research team showed that UPLC with MS could yield double the amount of peptides that were sequenced peptides as Nanoelectrospray . The signal-to-noise ratio showed a tenfold improvement in the UPLC-MS technique by at least a factor of 10 as compared to MS analysis (18).
- Another study in 2012 examined the bioequivalence of meldonium. Drugs applicable for an oral route (19).

Improvement in Resolving Power in Peptide Maps

The technique of peptide mapping is vital in the analysis of proteins. UPLC offers an added advantage of the analysis by causing a dip in the time required for the analysis and is also reportedly more sensitive and shows higher peaks with increase in resolution (20).

Pharmaceuticals examined by Multi-Residue Analysis

The water effluent of pharmaceutical companies was discovered to have levels of several molecules such as lipid regulators or psychiatric drugs.

UPLC coupled with Q-TOF-MS is used to verify and screen these drugs inside the samples of waste water remedy plant (21).

CASE STUDIES

Pharmaceutical analysis

The novel technique of Ultra-Performance Liquid Chromatography (UPLC) aids in reducing the time of the separations and the lesser amounts of solvents that are used. UPLC Acquity columns packed with UPLC BEH C_{18} with particles of dimensions 1.7 µm have been reported.

A team in 2006 checked several pharmaceutical formulations from an analysis of HPLC to UPLC to analyze the following:

- Hydrocortison that contained hydrocortisone acetate, methylparaben, propylparaben and hydrocortisone degradation product.
- Estrogel gel with estradiol, propylparaben ,methylparaben and estrone .
- Indomethacin possessing indomethacin and its degradation products 4-chlorobenzoic acid and 5-methoxy-2-methylindoleacetic acid and
- Triamcinolon with a mix of trimacinolone acetonide, methylparaben, propylparaben and triamcinolone.

The team reported that as compared to other columns used for analysis, the UPLC system could decrease the time for the analysis by a factor of nine when particles were used whose dimensions were 5 µm. A reduction of particle size to 3 µm in the columns caused a threefold reduction of time. The high pressure of 100 MPa with no adverse effects on the column (22).

Traditional Chinese Medicines (TCM)

The evaluation materials that constitute such traditional medicines by UPLC has been challenged by the presence of complex molecules in the samples. As TCM possesses varied levels of molecules that are complex, their characterization is challenged by the lack of a single molecule that contributes to its activity.

A challenging molecule of the Chinese Pharmacopeia has been Panax ginseng C.A. Mey. The molecules associated are very complex and include ginsenosides that are of two complex types: protopanaxadiols and protopanaxatriols. This causes a separation time that extends to hours. The application of UPLC to the characterization of the technique caused efficient separation of such metabolites with the same resolution albeit at lower times involved.

The advantage of UPLC over HPLC that has been reported is the separation akin to that of HPLC but with reduced time and solvent applied on the column (23).

Checking for drug abuse

A team reported the use of UPLC–MS/MS screening platform for the evaluation and quantification of generally used drugs of abuse in urine, i.e. a high-throughput quantitative analysis. Substances within the drug training opioids, stimulants and benzodiazepines and related retailers had been covered similarly to cannabis and pregabalin, a total of 35 unique analytes. Based at the concentrations and the physico-chemical homes of the substances, three parallel techniques were reported for UPLC–MS/MS. Prior to analysis, pattern instruction consisted of special simple dilutions with 60 and a hundred µL urine, respectively, using a Tecan Freedom Evo pipetting robotic platform. A Waters Xevo TQ-S tandem quadrupole mass spectrometer coupled to a Waters I-elegance UPLC was used for quantitative evaluation of one quantitative and one qualifying MRM transition for each analyte, except for tramadol for which the metabolite O-desmethyl-tramadol turned into covered inside the MRM approach to affirm tramadol identification. After development and initial trying out, the method has been efficaciously applied and automatically used at a clinic for quantitative screening of medication of abuse in extra than thirty thousand samples of urine (24)

Novel biomarkers for Epithelial Ovarian Cancer

Currently available checks are insufficient to differentiate sufferers with epithelial ovarian cancer (EOC) from controls. Plasma specimens of EOC patients and regular people had been analyzed the usage of UPLC/QTOF/MS. Eight biomarkers have been recognized which may additionally serve as novel biomarkers for diagnosis. In this take a look at, we discovered that the plasma metabolic profiling by means of UPLC/QTOF/MS finished a very good diagnostic accuracy in distinguishing EOC patients from regular controls, and the combination of the 8 selected biomarkers showed an brilliant class overall performance for the independent validation set (25).

Studies of drug toxicity

UPLC enables particular detection due to its high resolution during the drug development. Here, toxicity issue causes a cancellation of drug candidates and this reasons economic loss to the enterprise. It is a complex task to estimate candidate drugs for viable inhibition or initiation of metabolizing enzymes, toxicity or drug-drug interactions within the body. Further, its sensitivity also lets in the detection of the peaks at low concentrations. These factors reduce the time for evaluation and decrease failure of pattern analysis.

A 2015 study explored the toxicity of fluorochloridone in a mice study using UPLC to assess its effects (26).

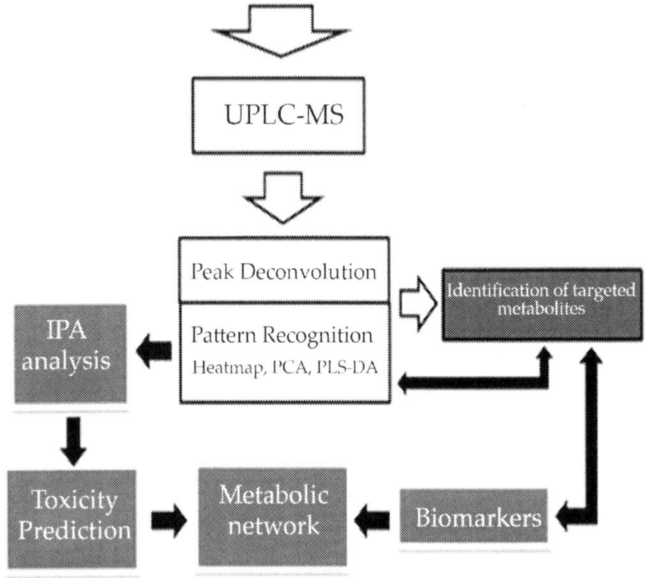

Figure 8.6: Workflow for the analysis of drug toxicity

Contaminants in food

Sudan I-IV is a red dye which is normally used in petrochemical industries. The in depth color of Sudan dyes lured frauds for improving the coloration of several spices and meals stuffs which can shape DNA adducts causing mutations. UPLC coupled to tandem mass spectrometry has aided identification of Sudan at low ppb degrees in spices and chilli containing meals stuffs (27).

Monitoring of drugs

Using the advantages of faster analysis: 5 minutes and 30s, a 2012 study evaluated the levels of β- lactam antibiotic in the blood of patients. The technique of UPLC-MS/MS was standardized for the quantifying of β-lactam antibiotics and β -lactamase inhibitors in plasma samples (28).

Explosives and UPLC

UPLC proved to be a more advantageous procedure to research various explosives. In addition, evaluation of explosive remains from hand swipes also can be detected through UPLC with very less time involved in the preparation of samples (much less than nine min), consequently enhancing the lab output as well as liberating up time consumed in an MS for additional characterization (29).

CONCLUSION

UPLC is one of the maximum important tools in analytical chemistry which will increase the velocity, resolution, and sensitivity of the chromatographic evaluation and reduces the time, solvent intake and cost involved.

The peaks obtained via UPLC have reduced noise and higher sign to noise ratio. It offers sharp and slender peaks of greater or less all categories of pharmaceutical tablets. It additionally helps the evaluation of complicated mixtures in much less time and the peaks obtained through this approach depicts more records that's clearer in contrast to the height obtained via HPLC.

UPLC is primarily based on a great deal demonstrated era of HPLC technique and it not only enhances however thrives the application of traditional HPLC method whilst many scientists commenced experiencing separation obstacles with it. The facts provided consistent with unit work through UPLC is tons more than HPLC because it has better speed, resolution and sensitivity. Also, the analysis time and solvent ate up in UPLC is less than in all formerly used chromatographic techniques. On the opposite, in UPLC the elements like resolution, tailing factors and peak area/height retention time repeatability are similar to those of HPLC.

REFERENCES

1. Kumar A, Saini G, Nair A, Sharma R. UPLC: A Preeminent Technique In Pharmaceutical Analysis. Acta Poloniae Pharmaceutica ñ Drug Research. 2012; 69 (3): 371-380.
2. Zhou C, Jin Y, Kenseth JR, Stella M, Wehmeyer KR, Heineman WR. Rapid pKa estimation using vacuum-assisted multiplexed capillary electrophoresis (VAMCE) with ultraviolet detection. J Pharm. Sci. 2005; 94: 576.
3. Swartz ME. Ultra performance liquid chromatography (UPLC): An introduction, separation science re-defined. *LCGC Suppl.* 2005; *8*: 8-14.
4. Lars, Y., Honore, H.S. On-line turbulent-flow chromatography-high-performance liquid chromatography-mass spectrometry for fast sample preparation and quantitation. *J Chromatogr.* 2003; *A 1020:* 59-67.
5. Jiang Z, Fisk RP, OíGara J, Walter TH, Wyndham KD. US Patent 6,686,035. (2004).
6. Acquity UPLC column. *https://www.waters.com/webassets/cms/library/docs/720001140en.pdf*
7. Wu N, Lippert JA, Lee ML. Practical aspects of ultrahigh pressure capillary liquid chromatography. J *Chromatogr. A.* 2001;*911*(1), 1-12.
8. Mezcua M, Agüera A, Lliberia JL, Cortés MA, Bagó B, Fernández-Alba AR. Application of ultra-performance liquid chromatography-tandem mass spectrometry to the analysis of priority pesticides in groundwater. J Chromatogr A. 2006; 1109(2): 222-227.

9. Cintron JM, Colon LA. Organo-silica nano-particles used in ultrahigh-pressure liquid chromatography. Analyst. 2002; 127 :701.
10. Martin M, Guiochon GJ. Effects of high pressure in liquid chromatography. Chromatogr A. 2005; 1090: 16.
11. Castro PJ, Plumb R, Granger JH, Beattie I, Joncour K, Wright A. Increasing throughput and information content for in vitro drug metabolism experiments using ultra-performance liquid chromatography coupled to a quadrupole time-of-flight mass spectrometer. Rapid Commun Mass Spectrom. 2005; 19: 843.
12. Zeng W, Han H, Tao Y, Yang L, Wang Z, Chen K. Identification of bio-active metabolites of gentiopicroside by UPLC/Q-TOF MS and NMR. Biomed Chromatogr. 2013;27(9), 1129-1136.
13. Tang YN, Pang YX, He XC, Zhang YZ, Zhang JY, Zhao, Z.Z.; Yi, T.; Chen, H.B. UPLC-QTOF-MS identification of metabolites in rat biosamples after oral administration of Dioscorea saponins: a comparative study. J Ethnopharmacol, 2015;165(165), 127-140.
14. Mao J, Xu Y, Lu B, Liu J, Hong G, Zhang Q, Sun S, Zhang J. Simultaneous determination of nicotine and its nine metabolites in rat blood utilizing microdialysis coupled with UPLC-tandem mass spectrometry for pharmacokinetic application. Anal Bioanal Chem. 2015;407(14), 4101-4109.
15. Lippert JA, Xin B, Wu N, Lee M. Separation of chiral pharmaceuticals using ultrahigh pressure liquid chromatography. Microcolumn. 1999; 11: 631.
16. Prakash L, Malipeddi H, Subbaiah BV, Lakka NS. Impurity profiling and a stability-indicating UPLC method development and validation for the estimation of related impurities of halobetasol propionate in halobetasol propionate 0.05% (w/w) cream. J Chromatogr Sci. 2015;53(1), 112-121
17. Karunakaran K, Navaneethan G, Elango KP. Development of a new RP-UPLC method for the determination of rabeprazole sodium in pharmaceutical formulation and application in dissolution studies. Trop J Pharm Res. 2011, 10(5), 655
18. Tolley L, Jorgenson JW, Moseley MA. Very high-pressure gradient LC/MS/MS. Anal Chem.2001 Jul 1;73(13):2985-91.

19. Pidpruzhnykov YV, Sabko VE, Iurchenko VV, Zupanets IA. UPLC-MS/MS method for bioequivalence study of oral drugs of meldonium. Biomed Chromatogr. 2012;26(5), 599-605.
20. Lauber MA, Koza SM, McCall SA, Alden BA, Iraneta PC, Fountain KJ. High-resolution peptide mapping separations with MS-friendly mobile phases and charge-surface-modified C18. Anal Chem. 2013;85(14), 6936-6944.
21. Petrovic M, Gros M, Barcelo D. Multi-residue analysis of pharmaceuticals in wastewater by ultra-performance liquid chromatography-quadrupole-time-of-flight mass spectrometry. J Chromatogr. A, 2006, 1124(1-2), 68-81.
22. Nováková L, Matysová L, Solich P. Advantages of application of UPLC in pharmaceutical analysis. Talanta. 2006; 68 (3) 908-918.
23. Yang GL, Yang LW, Li YX, Cao H, Zhou WL, Fang ZJ, Zhou HB, Mo JL, Xiao SX, Lin HR. Applications of ultra-performance liquid chromatography to traditional Chinese medicines. J Chromatogr Sci. 2010;48(1), 18-21.
24. Hegstad S, Hermansson S, Betnér I, Spigset O, Falch BM. Screening and quantitative determination of drugs of abuse in diluted urine by UPLC-MS/MS. J Chromatogr B Analyt Technol Biomed Life Sci. 2014; 947-948: 83-95.
25. Fan L, Zhang W, Yin M, Zhang T, Wu X, Zhang H, Sun M, Li Z, Hou Y, Zhou X, Lou G, Li K. Identification of metabolic biomarkers to diagnose epithelial ovarian cancer using a UPLC/QTOF/MS platform. Acta Oncol. 2012; *51*(4): 473-479.
26. Zhang S, Cheng X, Wang Y, Fan J, Li R, Zhou S, et al. Ninety-day toxicity and toxicokinetics of fluorochloridone after oral administration in rats. Int J Environ Res Public Health. 2015; 12(5): 4942-4966.
27. Schummer C, Sassel J, Bonenberger P, Moris G. Low-level detections of Sudan I, II, III and IV in spices and Chili-containing foodstuffs using UPLC-ESI-MS/MS. J Agric Food Chem. 2013;61(9). 2284-2289.

CHAPTER 9

FAST PROTEIN LIQUID CHROMATOGRAPHY

INTRODUCTION

FPLC is an acronym for Fast Protein Liquid Chromatography (1). GE Healthcare (formerly Pharmacia) coined the term FPLC and introduced to the Sweden market in 1982. It was originally referred to as fast-performance liquid chromatography as an opposite to high-performance liquid chromatography (HPLC, previously acronym for high-pressure liquid chromatography). Researchers working in laboratory use FPLC for preparative protein purification scale. To achieve desire purity, FPLC systems are useful in purifying proteins, peptides, and other biomolecules and yield in a consistent manner. By automation consistency could maintain in research process and also time saving to prepare for next step.

POINTS OF DISTINCTION BETWEEN FPLC AND HPLC

It is vital that both the techniques are distinguished: while FPLC finds application in the purification of proteins at a pressure and flow rate that is lesser, HPLC is employed for the purpose of analysis of molecules.

Now a days, FPLC systems are quite advanced that involve an enhanced application of automation. The use of automation bestows several advantages:

The parameters of running the procedure can be controlled that aids in the separation or purification of samples with minimum involvement of a controller.

Samples that display high sensitivity and those that show low expression are also resolved by the technique.

The use of a detection system aids in the appropriate downstream steps that include the collection of samples.

The optimization of protocols, experimental runs as well as analysis is aided.

PRINCIPLE OF FPLC

The technique is an extension of liquid chromatography for the analysis of resolution of mixtures. The separation in chromatographic techniques involves the use of differences in affinity a sample has for a stationary and mobile phase.

FPLC uses an aqueous mobile phase: the fluid used is maintained at a constant flow rate by the use of a pump. The composition of the fluid can be subjected to changes according to the requirements of an experiment.

A base that uses agarose cross linked to form beads is assembled in a column to be used as the stationary phase. The sizes and charges of the molecules used for the beads can be custom made.

The process in FPLC includes the use of a resin that facilitates ion exchange. A different buffer is used that finds application for the elution of samples that were initially bound to the column. The shift in buffer is achieved by a gradual shift from the binding buffer to the eluting buffer with caution exercised in the maintenance of a constant flow rate (2).

As samples are eluted from the column, samples are dissociated from the column to enter a detector. The detectors can be of several types that have been dealt with in Chapter 1: a common detection system is the use of the absorption spectrum of a protein in the UV range. Following detection, the samples are collected in appropriate fractions to be subjected to further studies such as MS (2, 3).

INSTRUMENTATION:

An FPLC instrument usually is composed of the following: A program controller, high-precision pumps, a mixer, a control unit, a column, a detection system and a fraction collector. The system can be subjected to manual operation but the use of automation or computers can aid the process.

The parts are summarized as follows:

Program controller: One of the vital aspects of a system, the controller facilitates the monitoring of the experimental protocol, the appropriate changes in various parameters apart from the collection volumes of the samples, data evaluation and the chromatogram development.

Pumps: The standard systems contain piston pumps that use 2 cylinders each applied for a buffer to be mixed for use. There are systems that operate on a peristaltic pump that pumps buffers from distinct containers to be combined in a separate mixing tank. The pumps can be operated at flow rates according to the experimental needs: whether the aim is analytical or preparative.

A Mixer: The apparatus aids in the mixing of sample buffers of varying grades of concentrations.

Column: Beads are packed into large tubes that serve as the column to function as the stationary phase in the separation. The column is held vertically within the system with the buffer traversing the column in the direction of gravity. The beads are kept in position in the column by the use of a frit that also permits the elution of the contents of the column. The samples are introduced into the column by the use of injectors and traverse the column with the aid of the mobile phase. The pressures employed in the system of FPLC is much less: 435-580 psi than what is applied in an HPLC system.

A variety of columns are available for separation using several principles such as ion -exchange or reverse phase or hydrophobic or size exclusion systems (4).

Injection loop: It refers to a piece of tube whose dimensions are known, the samples are loaded on this loop prior to their injection into the column.

Injection valve: It is a valve that serves as a connection between the injection loop and the mixer. Following application of a sample into the injection loop the valve facilitates the entry of the loaded sample into the chromatography column.

Flow Cells: these instruments evaluate the concentration of several samples in the eluant of the chromatography column. Apart from a UV absorption spectrum, the level of salts in the system is measured by conductivity measurements. These flow cells are linked to electronic systems for the amplification of signals generated.

Ultraviolet Monitor: The measurement of the absorption of samples at UV wavelength is achieved by this apparatus.

Monitor/Recorder: The instrument generates the output of the system to facilitate the visualization of samples that exit the column in the form of peaks on a chromatogram.

Fraction collector: It is a system of containers that rotate along an axis to permit the collection of fixed aliquots of samples as predetermined by the process controller (5,6).

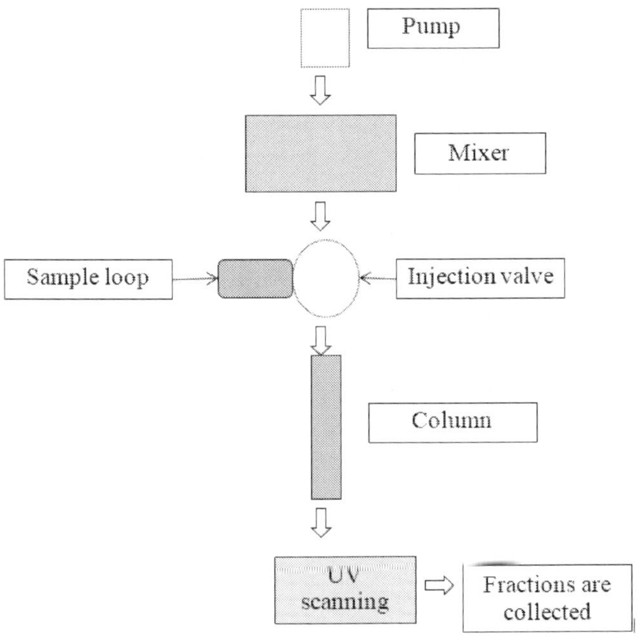

Figure 9.1: a schematic of FPLC

FPLC AND PROTEINS:

The technique of Fast protein liquid chromatography (FPLC) involves the separation of protein components. Using a combination of a resin as a stationary phase and a fluid/ liquid mobile phase, the separation or purification of proteins is achieved. The outcome of the experiment is reflective of the conditions used in the experiment that include the materials, resin as well as the efficiency. There are reports that any chromatography system so far has not yielded a product that is cent percent pure. A procedure that involves protein purification looks at achieving a desired product in a pure form with high yield in a rapid, economical and non-hazardous fashion. The protocols are designed in accordance with the source of the samples: natural sources include cell extracts or organisms while recombinant sources include vectors used in biotechnology.

The numbers of steps applied in a procedure determines the yield of a run in an FPLC system. Thus, careful optimization is necessary in order to carry out the system with minimum numbers of steps (7).

Ion exchange chromatography (IEC) finds application as an initial concentration step. Based on the type of particles used as stationary phase, small particles possess good resolving power while large particles can be applied for rapid flow rated. IEC can be applied to initially capture a sample prior to its subsequent use in a purification protocol.

Size exclusion chromatography is based on the retention time of a sample in a column owing to the packing of the particles. The technique offers advantages of less quantities of sample and the use of narrow columns for the separation.

Hydrophobic interaction chromatography (HIC): The principle of hydrophobic interactions governs the separation technique. The method is not affected by pH that encourages its use as initial or steps before the final yield (5).

APPLICATIONS OF FPLC:

Study of markers in the body fluids

FPLC has proven to be a fast and reliable technique for a study of markers in the body fluids such as plasma, urine, and cerebrospinal fluids which contain a variety of proteins and peptides. The study of the patterns of proteins in biological samples such as cerebrospinal fluids has been implicated as reflective of the state of health of an organism (8).

Other applications

As polynucleotides can also be studied by FPLC, rapid purification of RNA has been reported. (9, and 10).

In another application of FPLC, anion exchange chromatography using FPLC was used for the resolution of DNA from processed genetically modified (GM) oils. The purity of this isolated DNA was equivalent to that of cetyltrimethylammonium bromide method and High Pure GMO Sample Preparation Kit protocols (11).

To purify animal venoms, FPLC protein profiling has been used: A 2015 study used FLPC for separation of venom from *Rhopalurus junceus* (scorpion) venom (12).

It is used in identification of microvariability of a single protein. This study of microvariability is associated with outcomes that translate into clinical significance (13).

In analyzing the nitrogenous constituents of beer, which are believed to play a role in foam stability FPLC has been used fairly extensively (14).

THE APPLICATIONS OF DIFFERENT CHROMATOGRAPHIC TECHNIQUES USED IN FPLC

Ion Exchange Chromatography:

Based on their charge of ion, it separates mixtures. In this application, the negatively or positively charged functional groups can bind covalently to either anion or cation exchanger. Hydroxyapatite chromatography: For the purification and separation of proteins, viruses, enzymes, nucleic acids and other macromolecules Calcium phosphate ion $(Ca_3(PO4)_3(OH)_2$ can be used.

Affinity Chromatography:

A column consisting of a stationary packing material is bound to a particular ligand or substrate following which samples are applied. The sample components that possess an affinity for binding to the ligand in the column remain in the column by virtue of their interaction with the column. Elution of the bound components can be performed by the use of shifts in the ionic concentration or pH.

The use of affinity columns with Protein A find use in experiments involving antibodies. Following sample application on the column and washing to remove unbound components, the elution buffer can be applied in a reverse direction to facilitate the exit of the bound samples from the column.

Gel Filtration Chromatography:

The stationary phase is constituted by beads of known dimensions. When a sample is applied on this phase packed in a tube or column, bigger molecules are unable to enter the pores between the packing particles and are eluted from the column while smaller molecules can enter pores to elute out much later.

As the technique is based on separation of sizes that are excluded, the technique is called Size exclusion chromatography that finds use in the separation of several biomolecules, as reviewed by Lavanya et al (5)

Hydrophobic Interaction Chromatography:

Based on the hydrophobic patches on a sample, HIC uses the principle of hydrophobicity. Samples are applied in a HIC column using a buffer of high concentration of salts. (These salts facilitate the reduction of sample salvation). The decrease in salvation causes an expose of hydrophobic regions that bind the column to be eluted subsequently (15).

WHY USE FPLC OVER HPLC

From the name of fast protein liquid chromatography (FPLC) and high-performance liquid chromatography (HPLC), the differences between these chromatographic methods can be recognized. Extraction and purification of biomolecules is challenging; as they function in the temperature range within which organisms and human beings survive (16).

Protein samples are sensitive to parameters such as temperature extremes as well as the presence of organic solvents. These parameters alter the structure and functions of such samples that makes HPLC (that uses both these conditions) not an ideal candidate for proteins.

Keeping these concepts in mind, the technique of fast protein liquid chromatography was developed to provide a method for the separation of biomolecules.

Size: FPLC is aimed at the isolation of a protein samples in a form that is pure as well as unaltered. Hence, the technique is employed for the resolution and purification of several biomolecules of large sizes including DNA, RNA and proteins. HPLC is applicable in the analysis of samples of smaller sizes.

Volumes: As the purification of biomolecules involves the use of cell extracts, larger volumes of samples are permitted in FPLC than in analytical HPLC.

Column: The materials used for the columns in the two techniques are different. FLPC uses polymers or agarose particles of large sizes. As the technique uses lesser pressure, the resins are different and so are the column construction parameters. However, due to the high level of pressure applied in HPLC, the use of columns made of stainless steel and resistant to pressure is warranted. An advantage conferred by FPLC is as pressure is not a limiting factor in FPLC; the columns can be made of transparent materials. This facilitates the monitoring of the column through the experiment.

CASE STUDIES

FPLC vs. HPLC: separation of HbA2 in laboratory

Haemoglobinopathics are a group of autosomal recessive inherited human disorders pervasive in many parts of the world. The prevalence of β-thalassaemia in the lower portion of northern Thailand was found to be 1.4 percent and 24 percent haemoglobin HbE (17). For the prevention and control of severe β- thalassaemia disease: homozygous β-thalassaemia and β-thalassaemia with HbE (β-thal/HbE), heterozygote screening and genetic counselling are essential.

The diagnosis of β-thalassaemiais permitted by recording the measures of HbA2 values (an increase in levels) by the use of electrophoresis or measures of red blood cells or morphology studies of erythrocytes or high performance liquid chromatography (HPLC) (18). Quantitation of the levels of HbA2 using techniques of microcolumn chromatography (19) or cellulose acetate electrophoresis (18) are accurate along with high reproducibility, a major challenge is the use of skilled labour as well as larger duration of time for analysis.

The fast protein liquid chromatography (FPLC) method is a modification of microcolumn chromatographic techniques for HbA2 determination (19). The technique of FPLC has been applied for the separation of protein samples: the separation or purification. Hence, its use in the evaluation of analysis in a clinical application was daunting.

Certain studies however have recommended the use of FPLC for studies that involve variability in a protein molecule or the study of levels of proteins. These studies involved the resolution of several samples such as proteins from urine or plasma (20,22).

This prompted a team in 2009 to use FPLC for the resolution of HbA2 in addition to evaluate the method against conventional techniques of measurement such as HPLC.

The realm of clinical studies were not so far exposed to FPLC that required certain modifications. A column chromatography apparatus of micro dimensions was subjected to alterations in compositions of the buffer. It was reported that the system with a buffer of pH 8.1 yielded a good resolution of HbA2 and HbE. The protocol also displayed precise as well as reliable and accurate results. A procedure akin to that of HPLC was applied.

The column was packed with an anion exchange system: Hi Trap Sepharose that presents advantages such as the absence of abnormal behavior such as swelling or shrinkage or leakage. This permitted the analysis of a prodigious number of hemoglobin samples. An additional hallmark is the cost of a Hi Trapsepharose column was reported as 10 times lesser than that of a HPLC column.

The screening of HbA2 /E fraction was facilitated by a conversion of the chromatograms of this anion-exchange column into that of a conventional HPLC cation-exchange output. The data was subjected to quantitative analysis that showed a greater proportion of HbA2 in the technique of FPLC as compared to HPLC. The difference in values was attributed to the use of the weak anion exchange column. It was also reported that the rapidity of the system was on the lesser side as compared to HPLC and a compromise in the reproducible results was also marginally seen for this FLPC technique.

The chromatograms of the normal hemoglobin and β-thalassaemia sample were similar, the main difference being an elevated level of HbA2 to a marginal proportion. The elution profiles of HbE were comparable to the time reported for HbA2 of normal hemoglobin and β-thalassaemia samples. The main point of difference here was an increased profile of HbE as compared to HbA2 in the chromatogram. The levels and percentages of the chromatograms were done manually as the chromatograms displayed irregular baselines.

Though this irregularity can be a flip point of FPLC, nevertheless, the FPLC method served as an appropriate method for identification and quantification of the β-thalassemia as well as the levels of HbA2 /E.

This study suggested the use of an FPLC system in the analysis or screening of clinical samples apart from its role in purification and separation of samples as reviewed by S. Tangvarasittichai et al, 2009 (21). The following table shows the main comparison between FPLC and HPLC techniques.

Table 9.1: Comparison of main specifications of FPLC and HPLC:

	FPLC	HPLC
Aim	Purification	Analytics
Pressure	Max. 40 bar (4MPa)	1500 bar (150MPa)
Samples	Several kDa	×Å to approx. 3000Da
Sample volume	From 100 μL to several litres	100 μL
Eluent	Buffer	Solvents
Material hardware	Biocompatible	Stainless steel

Column material	Agarose polymer; only low pressure, big particle sizes, higher flow rates	Silica beads, pressure stable, small particle sizes, low flow rates
Methods	Size-exclusion, ion exchange, affinity or hydrophobic interaction	Reversed phase
Software	Volume-based method writing, intuitive, direct control	Time-based method writing, Analytical function

FPLC and Cultivars

For wheat cultivar identification, crude gliadin from common wheat was fractionated by rapid cation exchange chromatography (FPLC) on a mono-S column. Buffers which were used previously for gliadin fractionation on SP-trisacryl M (a soft cation exchange gel) were used first, so that it was thus possible to compare gliadin fractionations carried out under very similar conditions by cation exchange FPLC and low-pressure chromatography. In both of cases the elution times of the components were related to their charges at acidic pH. Fractionation can be achieved more rapidly on a mono-S column than on SP-trisacryl M, and the resolution is also greater. But the Mono-S FPLC did not yield pure gliadin component, each gliadin group was separated into several fractions. By the quantity of protein loaded onto the column, the elution times and the resolution can slightly be modified. However, such variations should not be critical in routine analysis of flour samples extracted under standard conditions. A well-controlled column washing procedure is necessary to ensure reproducible chromatographic profiles that can be used routinely to discriminate between common wheat cultivars. The 18 genotypes pattern were compared; all were distinguishable from each other. This preliminary investigation showed that cation exchange FPLC may provide an alternative method for wheat cultivar identification (22).

FPLC for plasma protein separation in biological samples: CSF and urine

The knowledge of the disturbances of plasma proteins, isoenzymes, hemoglobin, and the modulation of transport proteins due to disease has been built from research that depend on a wide range of techniques for protein separation. By combining immunochemical and electrophoretic techniques with ion-exchange and gel chromatography, it can resolve complex mixtures and reveal individual proteins structure with ever-increasing detail. However, such methods are too slow to be used as part of routine clinical chemistry. HPLC is a part of the routine equipment of many clinical laboratories for its advent of separating simple organic molecules. However, the rapid separation of proteins by HPLC has yet to be developed into a system that can provide high resolution over the wide range of protein types (enzymes, plasma proteins, and hemoglobin) required for the diagnosis and monitoring of disease. Preliminary reports on HPLC techniques led to the separation of blood and urine proteins through use of size-exclusion gels designed for HPLC (13)

The key to the novel FPLC system are the newly developed Monobead™ columns; very important in these three types of columns are the uniform monodisperse spheres, which give excellent separation and flow characteristics at only a moderately increased pressure (23). The types of produced columns are a strong cation exchanger (Mono STM), a strong anion exchanger (Mono QTM), and a column designed to provide properties similar to their PBE- 94 Polybuffer exchanger. It allows chromatofocusing over the pH range 9 to 4. This FPLC system and columns are designed to overwhelm the problems encountered by the separation of proteins by HPLC; which exploits many of the properties of proteins and their binding to resins.

Previous studies of the separation of urinary proteins and of several plasma proteins suggested that this form of chromatography might allow rapid identification of protein profiles or microvariability within a single protein, which could be of clinical significance (24). FPLC applied in the separation of urinary and cerebrospinal fluid (CSF) proteins, which were developed after optimizing buffer system on the anion exchanger Chromatography. The system was originally optimized for the separation of low-Mr proteins in the urine. By analyzing several causes of tubular proteinuria-acute pyelonephritis patients, severe burns patients, and renal transplants patients, the important features of the chromatograms looked constant, sharp peaks being produced by microglobulin, retinol-binding protein, eight repeated runs of the same urine to determine the CV of the peak position and peak heights, as measured by a Pharmacia LCC-500 peak analyzer aj-acid glycoprotein with a1-microglobulin. These features disappear when the lesion is reversible. The chromatographic trace of normal, unconcentrated urine has no peaks. However, due to the presence of large amounts of a wide range of plasma proteins in the urine, the patterns in mixed tubular and glomerular proteinuria could be readily interpreted.

These examples illustrate how FPLC is being used in the analysis of different forms of acute and chronic nephrotoxic damages (13). Urinary proteins, low-Mr proteins can be isolated from urine rapidly and, depending on their elution position, can be taken from the anion exchanger in relatively pure form. Preliminary studies of hemoglobin and its breakdown products in CSF suggest that FPLC could be a useful procedure in following the sequence of changes. The speed of separation and the reproducibility of the chromatograms obtained on the anion-exchange columns suggest that they could have an important role in profiling proteins in disease. The new anion exchange columns can be used for the rapid separation of monoclonal light chains under conditions that are unlikely to denature the protein and lead to their isolation within 10 mm of commencing the chromatography. After a simple salt exchange, the protein is ready for further studies such as chromate-focusing or tryptic digestion. The novel feature of FPLC is its speed as compared with other methods of investigating complex profiles of proteins. The anion-exchange system is especially valuable for separating plasma proteins, and chromate-focusing is useful for revealing their detail (13).

DRAWBACKS OF FPLC:

Some drawbacks of this technique are:

- Thc process cannot handle high pressures.
- FPLC warrants the use of glass columns for separation.
- The HPLC columns cannot be supported by FPLC instruments.
- It is extremely challenging to purify thermo liable (heat sensitive) proteins.

REFERENCES:

1. Amersham Pharmacia Biotech. AKTA FPLC System manual. (2000). Retrieved from http://www.hhmi.umbc.edu/toolkit/aktadesign.pdf.
2. Cutler, P. Protein Purification Protocols: Fast Protein Liquid

Chromatography. Totowa, NJ: Humana Press Inc. 2004.

3. David Sheehan, Siobhan O'Sullivan. Fast Protein Liquid Chromatography. Protein Purification Protocols. 2004; Volume 244: pp 253-258.
4. Madadlou A, O'Sullivan S, Sheehan D. Fast protein liquid chromatography. Methods Mol Biol. 2011; 681:439-47.
5. Lavanya, G., Mubarak, SK., Vishali, Sujana, K., Pramela, R., SyamaSundar, B. Protein Chromatography, Journal of Biomedical Sciences and Research. 2011; 3 (3) :424-438.
6. Wang, F., Min, Y., &Geng, X. Fast separations of intact proteins by liquid chromatography. J Sep Sci. 2012; 35(22): 3033-3045. doi: 10.1002/jssc.201200339.
7. David Sheehan. Fast Protein Liquid Chromatography (FPLC) Methods. Protein Purification Protocols. 1996; Volume 59.
8. Yin W, Carballo-Jane E, McLaren DG, et al. Plasma lipid profiling across species for the identification of optimal animal models of human dyslipidemia. *Journal of Lipid Research*. 2012;53(1):51-65. doi:10.1194/jlr.M019927.
9. Kim, I., McKenna, S. A., Viani Puglisi, E., & Puglisi, J. D. Rapid purification of RNAs using fast performance liquid chromatography (FPLC). RNA. 2007: 13(2): 289-294. doi: 10.1261/rna.342607.
10. Koubek J, Lin KF, Chen YR, Cheng RP, Huang JJT. Strong anion-exchange fast performance liquid chromatography as a versatile tool for preparation and purification of RNA produced by in vitro transcription. *RNA*. 2013;19(10):1449-1459. doi:10.1261/rna.038117.113.
11. Huang, L., Chen, Y., Wang, Y., Pan, J., & Tong, Y. Isolation of DNA from genetically modified oils by fast protein liquid chromatography. International Journal of Food Science & Technology. 2010; 45(7): 1495-1498. doi: 10.1111/j.1365-2621.2010.02295.x
12. Díaz-García A, Ruiz Fuentes JL, Yglesias-Rivera A, et al. Enzymatic analysis of venom from Cuban scorpion *Rhopalurus junceus*. *Journal of Venom Research*. 2015;6:11-18.
13. E. H. Cooper, R. Turner, E. A. Johns, H. Lindblom, and V. J. Britton.

Applications of Fast Protein Liquid Chromatography™ in the Separation of Plasma Proteins in Urine and Cerebrospinal Fluid. CLIN. CHEM. 1983; 29(9): 1635-1640.

14. C. J. Dale and T. W. Young. APPLICATIONS OF FAST PROTEIN LIQUID CHROMATOGRAPHY (FPLC) TO THE ANALYSIS OF THE NITROGENOUS CONSTITUENTS OF BEER. J. Insl. Brew. March-April. 1992; Vol. 98: 117-121.
15. McCue JT· Theory and use of hydrophobic interaction chromatography in protein purification applications. Methods Enzymol. 2009; 463:405-14.
16. Goke, B. &Keim, V. HPLC and FPLC. International Journal of Pancreatology. 1992; 11(2): 109-116.
17. Tangvarasittichai O, Sitthiworanan C, Dechgitvigrom W, Sanguansermsri T, Jeenapongsa R. Prevalence and heamatological parameters of thalassaemia in lower Northern Thailand. Haema 2005; 8 : 241-4.
18. Marengo-Rowe AJ. Rapid electrophoresis and quantitation of haemoglobins on cellulose acetate. J Clin Pathol 1965; 18 : 790-2.
19. Chamrasratanakorn T, Sanguansermsri T, Punyakeaw A. A modified microcolumn chromatography for HbA2 determination. Bull Chiang Mai Assoc Med Sci 1998; 31 : 32-5.
20. Shibasaki T, Gomi H, Ishimoto F, Miyahara T. Urinary N-acetyl-beta-D-glucosaminidase isoenzyme activity as measured by fast protein liquid chromatography in patients with nephrotic syndrome. Clin Chem 1990; 36 : 102-3.
21. S. Tangvarasittichai, O. Tangvarasittichai & N. Jermnim. Comparison of fast protein liquid chromatography (FPLC) with HPLC, electrophoresis & microcolumn chromatography techniques for the diagnosis of β-thalassaemia. Indian J Med Res. 2009; 129: 242-248.
22. C.Larre, Y.Popineau, W.Loisel. Fractionation of gliadins from common wheat by cation exchange FPLC. Journal of Cereal Science1991; 14 (3): 231-241.
23. Soderberg L, Bergstrom J, Andersson K, et al. Physico-chemical considerations in the use of Monobeads for the separation of

biological molecules. Protides Biol Fluids Proc Colloq. 1983; 30: 629-634.

24. Wahlstrom L, Muller RM, Fagerstam L, et al. Two-dimensional analysis of human sera by chromatofocusing on Mono P followed by fused rocket immunoelectrophoresis. Protides Biol Fluids Proc Colioq. 1983; 30: 653-655.

CHAPTER 10

ION EXCHANGE CHROMATOGRAPHY

INTRODUCTION

It is a process that allows the separation of ions and polar molecules based on the charge properties of the molecules. The solution that is analyzed is a *sample*, and the fractions that have been separated are called *analytes*. The technique can be applied for the resolution of molecules that bear a particular charge such as biomolecules (1).

PRINCIPLE

Ion Exchange Chromatography relies on charge-charge interactions between the proteins. Ion exchange chromatography retains analyte molecules based on ionic interactions.

The sample molecules interact with functional groups that are attached on the surface of a stationary phase: the charges of both are opposite.

Mechanism of ion-exchange chromatography or Selectivity in ion-exchange chromatography may be de- fined as the relativity of sample ions to form an ion pair with the sites of opposite charge within the stationary phase.

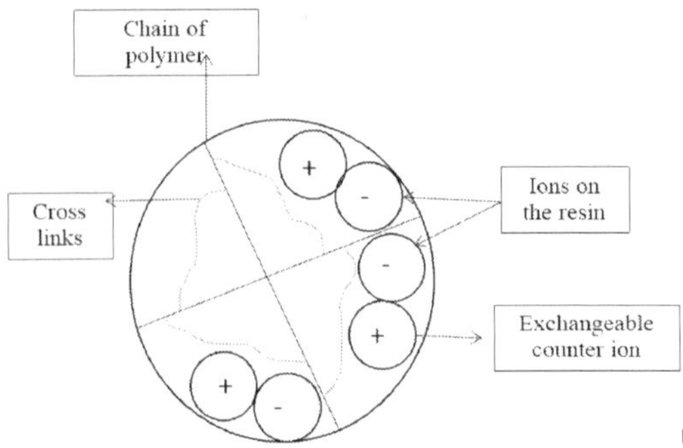

Figure 10.1: Set up of an IEC experiment

The selectivity is affected by the following two parameters:

(1) Electrostatic attraction (ES): This attraction is stronger when the sample ion has a higher charge and when the paired ions are very close to one another.

(2) Enforced pairing effects (EP): Stronger ion pairing occurs within the stationary phase than in the aqueous solution due to a combination of effects that may include hydrophobic attraction, hydrogen bonding, lower dielectric constant and what has been termed "water-structure induced ion pairing.". Large, poorly hydrated univalent ions in aqueous solution intrude into the surrounding water structure without being able to strongly orient the water molecules around themselves into coordinate hydration shells. This contributes to a tightening of the water structure around the ions. The chemical results are such that large ions tend to be rejected by the water phase and they are more easily extracted by nonstructured organic solvents than are smaller ions. The strong retention of polarizable inorganic and organic anions in ion-exchange chromatography is likely the result of EP effects (1,2).

The exchange reaction in IC can be written: $RN{+}E{\cdot}{+}{\cdot}I \longrightarrow RN\text{-}I{\cdot}{+}{\cdot}E$

where E– The ion in the eluent

I– Ion in the analyte

RN- the functional group

He constant RN+ sites at the ion exchanger result in a strong electrostatic "wall" that excludes cations from the cell segment from entering the stationary section.

For example, an anion-exchange column prepared by coating a reversed-phase HPLC column with a cationic surfactant has a positive zeta potential. This facilitates the approach of anions from the mobile phase to the mobile-stationary phase interface and to close proximity of the exchange sites. Since the coated stationary phase contains only positive sites fixed to a hydrophobic matrix, cations from the mobile phase would be excluded.

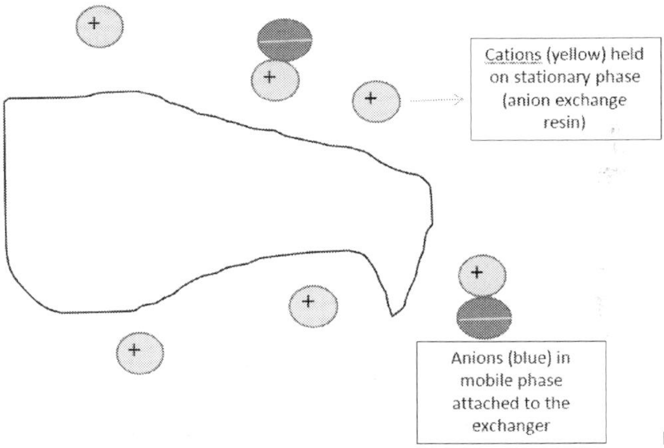

Figure 10.2: Scheme of attraction between ions

The physical nature of the ion exchangers used in modern IC is also conducive to exchange of sample and eluent ions. For example, the most widely used anion exchangers have a thin coating of charged latex particles coated onto the outer surface of a spherical substrate. The line charge density of the latex polymer structure as well as the favorable geometry favors the electrostatic component of the ion excanger mechanism. A small ion with a 2− charge, such as sulfate, ranks high in the selectivity series in IC due to this strong electrostatic component.

TECHNIQUE

This technique uses two materials that include phases that do not mix with each other: a mobile phase that traverses a column unidirectionally and a stationary phase allows separation of molecules (1-3).

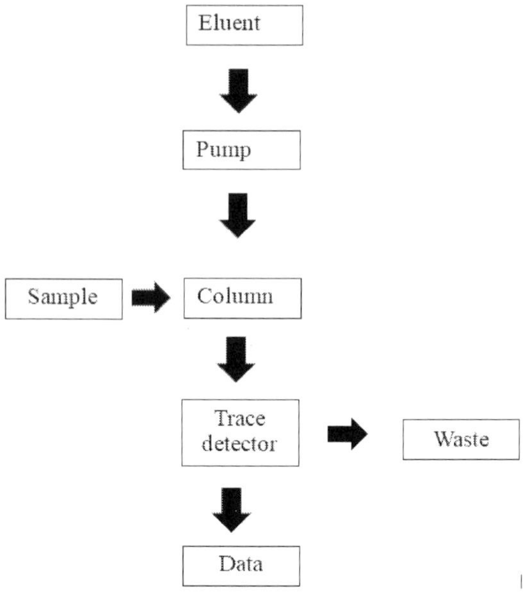

Figure 10.3: Instrumentation

The system is composed of a pump followed by a sample injector, columns, detectors and and a data system.

Pump

The pump in the instrumentation ensures an uninterrupted flow of solvent along the entire system. The most practical system for the delivery of the mobile phase is that which can combine several liquids in different proportions at the command of the operator. There is a series of mobile phase reservoirs that can contain a range of different mobile phases that can be used individually, blended or for mobile phase programming purposes "gradient elution".

In general liquid chromatography, the reservoirs can be stainless steel but in ion chromatography in which the mobile levels may have severe pH values, the reservoirs want to be manufactured from glass or ideally a suitable plastic together with PEEK (polyether-ether-ketone). The advantage of PEEK is that it's also inert to many organic solvents that could want to be used inside the mobile phase. In truth, all additives of an ion chromatograph which can are available in contact with both phase of the distribution device ought to be comprised of suitable inert fabric. This includes all cellular phase conduits, valves, pumps, sampling devices, columns, detector sensor cells, etc.

The solvent reservoirs are connected to a solvent choice valve and a solvent programmer wherein a specific solvent or particular solvent application can be selected. The solvent then traverses to a high efficiency pump. The cell segment passes from the pump to the sampling device, generally an easy rotating valve that on rotation places the pattern in line with the cell waft which then passes onto the column.

The *go out waft* from the column passes either to an ion suppressor (if you want to be discussed later) or without delay to the detector. Gas may also come out of the answer on the column exit or inside the detector, ensuing in sharp spikes. Spikes are created through microscopic bubbles which exchange the nature of the flowing circulation making it heterogeneous. The waft may additionally arise as those microscopic bubbles regularly amassed and mixed in the detector mobile. The first-rate outcomes can be received by way of applying vacuum to each solvent for about 5 min.

Pumps types

The *constant-flow pump* is the most widely used in all common IC applications. Flow rate stability is an important pump feature that distinguishes pumps. For size exclusion chromatography, the flow rate has to be extremely stable. External electronic control is a very desirable feature when automation or electronically controlled gradients are to be run.

Constant flow pumps

Constant-flow systems as how two variations: reciprocating piston and positive displacement pumps. Reciprocating piston pump can maintain a liquid flow for indefinitely long time.

Reciprocating piston pumps

A retraction of pistons alters the pumping rate. The development of pulses as sine waves poses a challenge that warrants the use of dampeners.

Dual piston pumps

It supplies a non-fluctuating stream. Both pump chambers are driven by the same motor through a common eccentric cam; this common drive allows one piston to pump while the other is refilling.

The system offers several advantages such as: a larger flow rate and a rapid rate of changing while a challenge is a compromise in reliability in case of alterations in the mobile phase.

Injector

Sample introduction can be accomplished in various ways. The simplest method is to use an injection valve. In more sophisticated LC, automatic sampling devices are incorporated where sample introduction is done with the help of auto-samplers and microprocessors (4).

In liquid chromatography, liquid samples may be injected directly and solid samples need only to be dissolved in an appropriate solvent. It is not mandatory for the solvent to be associated in the mobile phase but is preferred to avoid interference issues. It is always best to remove particles from the sample by filtering, or centrifuging since continuous injection of particulate materials will eventually cause blockage of injection devices or columns.

Injectors have to provide the opportunity of injecting the liquid pattern in the range of 0.1 to one hundred ml of quantity with high reproducibility and beneath excessive pressure (as much as the 4000 psi). They ought to additionally produce minimal band broadening and decrease viable drift disturbances.

The maximum beneficial and extensively used sampling device for present day LC is the microsampling injector valve. With these sampling valves, samples may be added reproducibly into pressurized columns without big interruption of glide, even at increased temperatures. With commercially available automated sampling gadgets, big numbers of samples may be mechanically analyzed by means of LC without operator intervention. Such system is famous for the evaluation of ordinary samples (quality manipulate of drugs), in particular whilst coupled with automated records-coping with systems.

Automatic injectors are fundamental in for chromatographic parameters including solvent selectivity, and temperature optimization. Most of the autosamplers have a piston metering syringe type pump to suck the pre-established pattern volume into a line and then switch it to the fairly big loop (~100 ml) in a standard six-port valve. The only autosamplers make use of the unique vials with pressurization caps. A special plunger with a needle, push the cap right down to the vial and displace the sample through the needle into the valve loop (4).

Columns

The principle of ion exchange chromatography is that, charged molecules bind electrostatically to oppositely charged groups that have been bound covalently on the matrix. Ion exchange chromatography is a type of adsorption chromatography so that, charged molecules adsorb to ion exchangers reversibly so, the molecules can be bounded or eluted by changing the ionic environment. Ion exchangers can be used in column chromatography to separate molecules according to charge; actually, other features of the molecule are usually important so that the chromatographic behavior is sensitive to the charge density, charge distribution, and the size of the molecule.

Figure 10.4: A Metrohm 850 Ion chromatography system

An ion exchanger is usually a three-dimensional network or matrix that contains covalently liked charged groups. If a group is negatively charged, it will exchange positively charged ions and is a cation exchanger. An example cation exchanger is carboxy-methyl group. A positively charged material, will exchange negatively charged ions and is an anion exchanger. Di-ethyl-amino-ethyl group (DEAE) is an example of a group used in anion exchanger. The separation on an ion exchanger is usually accomplished in two stages:

The sample components are encouraged to bind the exchanger.

Elution of the column is performed with shifts in pH or ionic strength

The use of an anion exchanger is suggested, if a protein is stable at pH value above the isoelectric point, while a cation exchanger is required for below the isoelectric point. The columns used may be stainless steel, titanium, glass or a non-reactive plastic. The dimensions of the column may be between 3cm up to 50 cm in length and up to 5 cm in diameter depending on the application. The samples dictate the time a column can be used (5,6).

Guard column

It serves for protection of the column when arranged in front of the column. They eliminate materials that could block the main separation column that has the potential to lead to false results or a dip in sensitivity as well as resolution.

Suppressor

The reduction in conduction of substances that are applied for the elution of analytes is performed by the suppressor. These devices are based on membranes and remove such conductivity of the surroundings. They are desalting devices that eliminate ionic salts present in the eluent that could cause interference.

Prior to the experiment, it should be ascertained whether the suppressors must be applied for the eluents in the experiment: either water based or organic. The experiments involving such eluents can use suppressors that are:

- Chemical (micromembrane suppressor)
- Electrolytic (self-regenerating suppressor)

Suppressors protonate analytes that are weakly acidic that could lead to the precipate or adsorption on the basis of hydrophobicity. Suppressors deprotonate analytes that are weakly basic to form species with no charge.

The micromembrane suppressors are composed of layers of ion-exchange membranes along with a chamber screen converted to have a high potential for exchanging ions. A chance of an analyte entering across the suppressor membranes that could lead to lack of detection has also been reported for such suppressor systems.

The latter is applied more on a commercial basis due to its autonomous operation and the potential of use of water as raw material. However, the current applied should be carefully supervised to circumvent potential damage and also the prevention of oxidation products of organic eluents that could yield ionic products. The suppressor may also adsorb or precipitate certain analytes that could lead to their lack of detection.

Detectors

A detector in order to be applied successfully on a run can possess following features that include:

- Sensitivity
- Minimum dead volume
- Rapid response
- Broad range
- Minimum noise
- Reliable
- Simple and should not destroy sample analytes (5,6)

The following types of detectors are applied:

Electrical conductivity detector:

The instrument is composed of 2 electrodes in a flow cell made of glass. As the ions enter the flow cell, a difference in signal is generated that is read by a circuit to be related to the concentration of the species. This is followed by a digitalization of the output to be read by potentiometry.

Amperometric detection:

The technique has high sensitivity and is applied for analytes that can be easily modified by oxidation or reduction. A variation called Pulsed Amperometric Detection (PAD) finds application in the detection of carbohydrates using anion exchange.

Mass Spectrometry:

Using techniques of electrospray, atmospheric pressure chemical or photo ionization, MALDI the detector classifies ions on the basis of the Mass to charge ratio that is highly specific.

Sensitivity of the Detector is a vital component of a separation as the discrimination between n analyte and potential false is a huge task. The reduction in peak sizes has been associated with a necessity of absence of baseline readings without noise.

Data System

An accurate interpretation of the data generated minimizes the involvement of an operator. An integrator that is programmed earlier can be used for simple setups while a small operating system may be necessitated when large scale studies are performed (Yasser et al, 2013, Jungbauer A and Hahn R, 2009).

TYPES OF IEC

There are two main classes of IEC:

Anion exchangers: They are molecules with a negative charge based on a support that bears positive charge. Examples include:

Q-resin, a Quaternary amine; and DEAE resin, DiEthylAminoEthane

Cation exchangers: --- They are molecules with a positive charge based on a support that bears Negative charge. S-resin, CM s etc. are examples of cation exchange resins (2).

RESINS USED

An ideal matrix for use in an IEC experiment should have the following features:

- Large amount of porosity.
- Chemically inert
- Resistant to changes in pH or ionic strength
- Stability
- Does not facilitate general binding.
- (Harvard Apparatus)

The resins used are small beads with densities greater than water and are contained within the columns. They serve to filter material and achieve the desired resolutions. They are composed of an insoluble matrix to which charged molecules are attached. Weak exchangers include:

- Carboxymethylcellulose (CM-cellulose) that have the functional group: CH2OCH2COOHthat bears a negative charge upon ionization to serve as a cation exchanger.
- Diethylaminoethyl-cellulose (DEAE-cellulose) bears a positive charge to find use as an anion exchanger.
- Sepharose materials are applied for the separation of samples with high molecular weight. (4).

Additional types include:

- Gel type strongly acidic cation resins are composed of Styrene and DVB (divinyl benzene) with the presence of sulphonic groups.
- Gel strongly basic anion exchange resin are composed of Styrene and DVB with presence of quaternary amine active groups,
- Porous type strongly acidic cation resins. Have a similar composition but are macro porous physical structure to facilitate adsorption and elution of organic substances.

- Porous type strongly basic anion exchange resins. are composed of Styrene and DVB with presence of quaternary amine active groups with porosity to facilitate adsorption and elution of organic substances.
- Weakly acidic cation exchange resins are composed of Acrylate and DVB polymerization with carboxylic active groups, with porosity facilitate adsorption and elution of organic substances.
- Weakly basic anion exchange resin are composed of Styrene and DVB polymerization, with tertiary amine active groups, as reviewed by Mohammed JS, 2017 (5)

A REPRESENTATIVE PROCEDURE OF IEC

- Using a manual or autosampling method, samples are applied into a sample injector.
- A buffered solvent is applied that facilitates the flow of sample across the stationary phase.
- Sample molecules that bind the stationary phase can be subjected to elution using a change in ionic strength that helps in the replacement of adsorbed ions.
- This is followed by detection

ELUTION OF BOUND COMPONENTS

Bound protein analytes can be eluted (desorption) in one among two feasible ways:

- pH;
- Ionic strength.

A change in pH alters the net charge of the certain protein, and for this reason its matrix binding capacity. More usually, increasing the attention of a similarly charged species inside the cell segment can compete with and in the long run displace the bound ionic species.

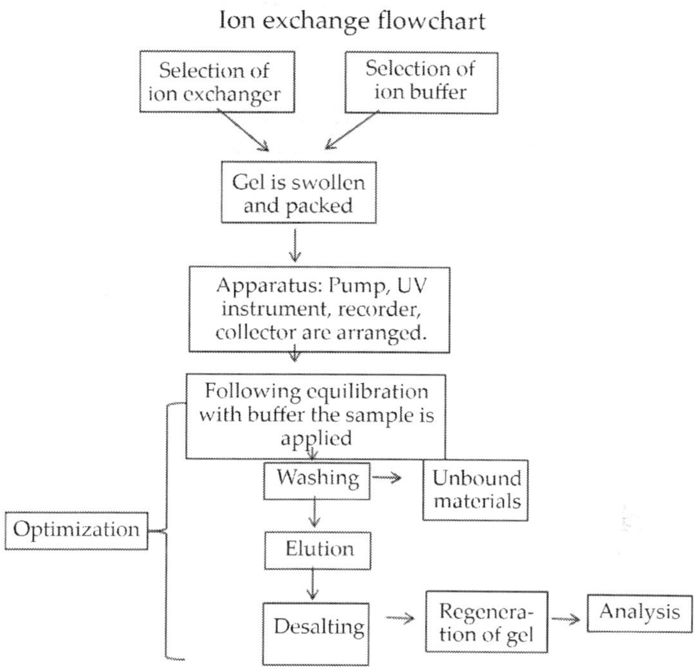

Figure 10.5: Scheme of an IEC experiment

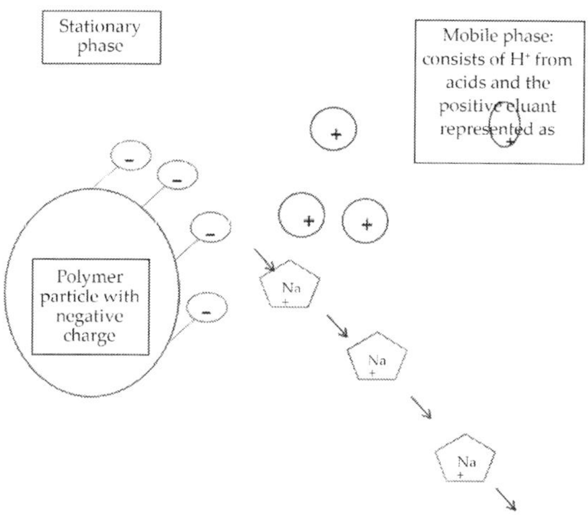

Figure 10.6: Scheme of elution: A replacement of exchange of ions here in the figure sodium ions yields the separation of the proteins/analytes bound to a column.

During anion-exchange chromatography for instance, negatively charged protein analytes may be competitively displaced by means of the addition of negatively charged chloride ions (e.g. from sodium chloride). By steadily increasing the salt levels in the cellular phase, the affinity of interplay among the salt ions and the functional businesses will eventually exceed that which exists between the protein fees and the practical agencies, ensuing in protein displacement and elution.

Mobile phase pH may be altered to want both adsorption and elution of proteins. In standard, a pH is chosen to simply allow binding of the goal protein. This is commonly about 1 Ph unit above or underneath the target protein pI. A extra distinction in pH might lead to more potent protein binding. A pH change also can be used to set off desorption of the target protein (a pH lower within the case of anion-exchangers and vice-versa).

As with pH, ionic alterations can also be used to manipulate target protein adsorption and elution. As a widespread rule, the highest ionic power with a view to permit adsorption and the bottom ionic power with the intention to allow elution are endorsed (6).

FACTORS FOR IEC

Samples and Sample Buffer

The parameters of samples and their buffers should be in similar states to avoid the formation of artifacts such as bubbles. Washing of the columns is mandated following sample application to minimize the level of non-specific binding and interactions. An increase in the ionic strength of buffer causes an elution of analytes that bind the column.

Selectivity

The success of an experiment depends on resolution that in turn depends on the nature and quantity of functional groups present, the parameters such as Ionic strength, pH and elution.

Elution can be achieved in a:

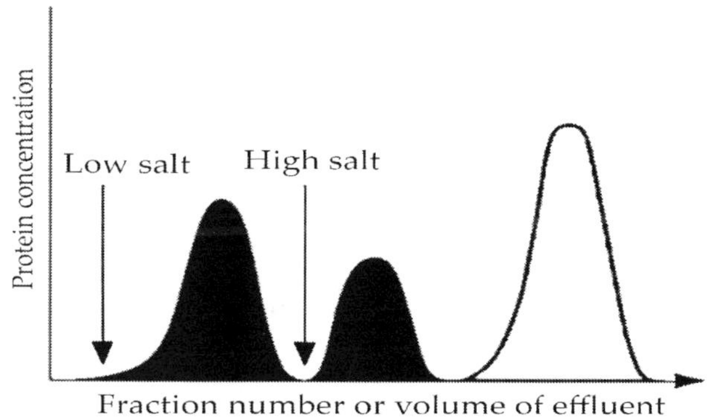

Figure 10.7: Chromatogram showing the peaks of elution at different salt concentrations

1. *Linear Gradient or Step Elution Specific*: Proteins are eluted by means of a boosting the ionic power of a buffer. The UV and conductivity will display the elution peak and adjustments in salt attention.
2. *Gradient Elution*: When beginning with an unknown pattern and for excessive separation evaluation.
3. *Step Elution*: When the ion alternate separation has been optimized the usage of gradient elution. This accelerates the separation time.
4. *Group Elution*: Group elution is likewise a Step elution. It is applied for the removal of unwanted substances or impurities.

The list of buffers used is summarized in the following table:

Table 10.1 Buffers Used In IEC: (Reported for concentrations of 20mM)

Buffer	Type	Anion	pKa	Region of activity
L-Histidine	Anion exchange buffer	Cl-	6.15	5.5-6.8
Bis-Tris			6.5	5.8-7.0
Bis-Tris propane			6.8	6.4-7.3
Triethanolamine			7.7	7.3-8.2
Tris			8.16	7.5-8.8

Formate	Cation exchange buffer	Na+	3.75	3.3-4.3
Acetate			4.76	4.2-5.2
MES			6.15	5.5-6.7
HEPES			7.55	7.6-8.2

ION EXCHANGE: WHY TO USE?

Depending on the extent of ionization of functional groups present on an exchanger there are either:

Strong ion exchanges whose ion exchange capacity is not affected with pH changes. They are recommended when:

Samples do not interact to yield intermediate forms. As the charges are not affected, the separation techniques and optimization is rapid. Weak ion exchanges vary with changes in the pH. They present a wide range of separations that can be performed and higher selectivity.

APPLICATIONS OF IEC

IEC in drugs:

Since, 1950 the technique of ion exchange has been applied to delivery of drugs or masking of taste as they offer advantages such as: a reduced frequency in administering a drug, as well as a decrease in the fluctuations and accumulation of the administered drug. The technique has also been reported to be economical.

Currently in the market, amphetamine on a cation resin: has been applied for control of appetite by Strasenburgh Laboratories.

Other products also include a flu and cough relief called Penntuss which is a combination of Codeine and Chlorpheniramine and Ionamin (phentermine) (7).

Research in drugs:

The field of drug research has explored the potential of IEC. The analysis of compounds that possess several active components is a challenging task. A polar molecule is expected to be retained in an IEC column to achieve both separation of polar and non-polar molecules. A 2009 study showed the efficiency of IEC in the separation of several important drugs that included: metformin HCl, isoniazid, sodium nitroprusside, timolol maleate, naproxen sodium, and clonidine hydrochloride (9).

The studies of dissolution of drugs have been evaluated using IEC. The analysis and dissolution studies of aminoglycoside antibiotics was tested using Ion chromatography (9). Another study evaluated the application of IEC in the analysis of driugs and products for impurieties or unwanted components (10).

For separating molecules

IEXC is one of the extra effective protein purification techniques available and likely the most regularly used chromatographic approach for the separation of proteins, polypeptides, nucleic acids, polynucleotides, and different charged biomolecules. A major plus of this approach is that the elution typically takes area under mild conditions, so that the protein can preserve its local conformation in the course of the chromatographic method.

In general, ion exchangers are more densely substituted than different adsorbents used in protein chromatography and its potential for protein binding is very excessive.

Proteins are molecules with both positive and negative charges on their surfaces. The pH at which the net charge is zero is the isoelectric point (pI) that depends on the charged amino acids. Positive charges are reflected by the presence of arginines, lysines and histidines while negative charges are supplied by means of aspartate and glutamate residues and the C-terminal carboxyl organization.

Generally, anion trade chromatography (AIEXC) is done at pH values above the isoelectric point of the protein under study, while cation alternate chromatography (CIEXC) is accomplished beneath the isoelectric factor.

Most proteins are negatively charged at physiological pH values (pH 6 – 8) and consequently, in many applications of IEXC a first approach is to use an anion exchanger. A distribution profile of the isoelectric factors of a wild-kind E. Coli proteome showed that 95% of the intracellular proteins had been negatively charged at fundamental pH and need to consequently maximum effortlessly be eliminated the use of CIEXC.

Typically, IEXC is used to bind an analyte and then wash away non-bound contaminants, as reviewed by Karlsson, 1998 and Hedhammar (11, 12)

CASE STUDIES

In studies of small-angle X-ray scattering

Many purification protocols involve the use of ion-exchange chromatography for the separation of macromolecules. A 2016 study applied a technique of ion-exchange chromatography directly before small-angle X-ray scattering studies.

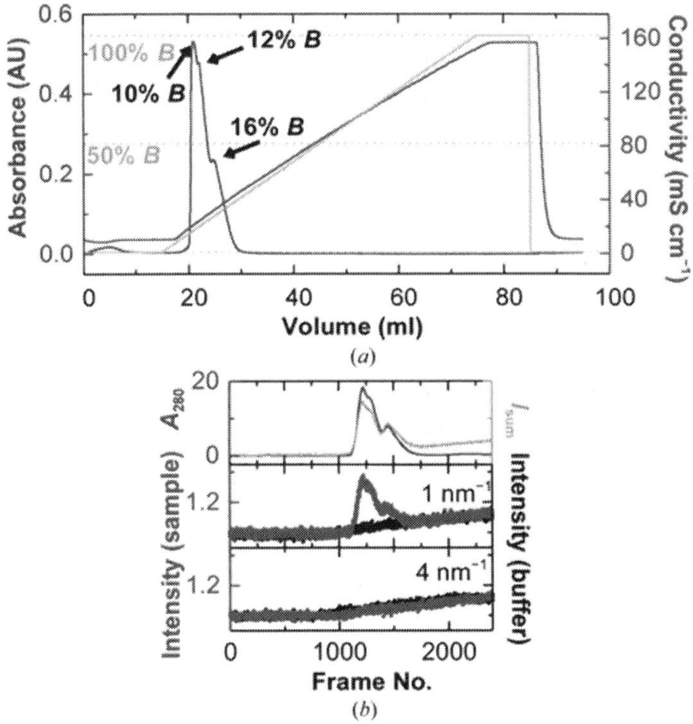

Figure 10.8: Linear gradient IEC–SAXS performed on BSA. (a) Standard ion-exchange chromatogram of BSA on the Uno Q-1R column, highlighting the locations of the three peaks. Orange line, percentage of buffer B; red line, resulting conductivity; violet line, absorbance at 280 nm. (b) IEC–SAXS chromatograms of BSA using a linear salt gradient. Top panel, UV absorbance (violet) and total scattering intensity (green). Middle and bottom panels, chro-

matograms of sample (red) and buffer (black) runs at q = 1 nm−1(middle) and 4 nm−1 (bottom) (13).

Using a linear gradient protocol, the need for offline tests is obliterated. The background correction systems using an aligned run of a buffer avoided the use of interpolation. By a cautious selection of the salt in the step-elution method, this method permits the separation of the peaks to be stepped forward, as no limit exists at the quantity of sub steps.

Online ion-alternate chromatography provides some other critical biochemical purification technique to the repertoire of purification methods which may be coupled with SAXS. Proteins of comparable apparent molecular weight and a variation in charges can be separated using this application.

For IEC, buffer conditions, which include salt concentrations or pH, are commonly determined by the biochemical characteristics of the analyte under study, which includes its isoelectric point and solubility. It has been recommended that reduction in the amounts of all additives for maximization of the X-ray comparison among the protein and the buffer in a SAXS experiment (13)

Determination of isotopes

With a view of determining the compound-specific nitrogen isotopic compositions of amino acids a team in 210 used cation-exchange chromatography. This step was applied before derivatization for gas chromatographic separation. It was reported that the recovery rates of amino acids were above 90%/ The applicaion of techniques to several samples from environment such as: calcareous, siliceous, rock and sediment samples along with aggregated microbial samples and biological soft tissue samples.

It was concluded that cation-exchange chromatography is a valuable technique for be application determination of isotopes of amino acids (14).

Ion Exchange Treatment of Drinking Water

Ion alternate (IE) is a water remedy method in which one or extra undesirable contaminants are eliminated from water. If the contaminant has a plus charge, it'd be known as a cation, and could be removed through use of an IE media called a cation alternate resin. If the contaminant has a minus charge, it'd be called an anion, and the correct treatment media could be called an anion trade resin.

A system of tanks is recommended for the removal of contaminants: The smaller tank carries the sodium or potassium salt used to regenerate the resin media while the taller tank(s) carries the purifying media called a "cation" change resin. As water traverses the column, divalent ions such as calcium, magnesium are exchanged for sodium or potassium ions in the resin. With use of the resin, depletion of the ions occurs that can be replenished by an application of salt in aqueous solutions (15)

Ion exchange in food industry

The following applications are a glimpse of the applications of ion exchange in the food industry:

The process of removal of mineral cations and amino acids: demineralization is essential in the preparation of grape sugar in order to be used as a sweetener and production of wine.

The wine making industry requires the exchange of polyphenols in fruit samples with organic acids that are then used in wine making.

The application of ion exchange columns in the dairy industry involves exchange of mineral cations and amino acids by a cation resin and the anions and organic acids by a special anion resin with a minimum loss of protein for demineralized whey that can be consumed by infants.

Removal of bitterness from Orange juice involves the use of adsorbent resins. In order to avoid the addition of sweeteners, an adsorbent resin can remove compounds such as Limonine, Naringine without altering properties of the juice.

A SUMMARY

The technique of ion exchange uses principle of separation of molecules by their binding to a material of opposite charge on a column. The advantages of the technique include:

- Rapid separations
- The resolution is high.
- The samples have been recovered with high proportions.

- The buffers used do not denature materials and can be extended for subsequent processes.
- The concentration of a sample: recovery of analytes from dilute systems can be achieved

The challenges in the technique of ion exchange include:

- An initial binding of samples to a column requires a condition of altered pH or ionic strength than when the samples are eluted.
- The instruments require resistance to ionic conditions applied.
- The technique can yield products with elevated levels of salt.

REFERENCES

1. Yasser M. Moustafa and Rania E. Morsi. Ion Exchange Chromatography - An Overview, Column Chromatography, Dr. Dean Martin (Ed.), InTech. 2013: DOI: 10.5772/55652. Available from: https://www.intechopen.com/books/column-chromatography/ion-exchange-chromatography-an-overview.
2. Jungbauer A, Hahn R. Ion-exchange chromatography. Methods Enzymol.2009; 463:349-71. doi: 10.1016/S0076-6879(09)63022-6.
3. Guide to Ion-Exchange Chromatography. Published by Harvard Apparatus.
4. Scopes, RK. Ion exchangers-principles, properties and uses. In "Protein Purification: Principles and Practice". 1982: pp75-101. Springer-Verlag, New York.
5. Mohammed JS; A brief review on Ion Exchange Chromatography; PharmaTutor; 2017; 5(2); 30-38.
6. Philip M. Cumminsa, Oonagh Dowlinga and Brendan F. O'Connora. Ion-Exchange Chromatography: Basic Principles and Application to the Partial Purification of Soluble Mammalian Prolyl Oligopeptidase.
7. S. Sivaneswari, D. Veena, P. Sai Sumana, P. Subhashree, L. Ramya, R. Rajalakshmi, P. J. Chandana and E. Karthikeyan. Ion exchange resins as drug delivery carriers. Journal of Chemical and Pharmaceutical Research, 2015, 7(4):1436-1445.

8. Rogers LA, Crews KE, Long SG, Patterson KM, McCune JE. Evaluation of Chromatographic Methods for Drug Products Containing Polar and Non-Polar Molecules Using Reversed Phase, Hydrophilic Interaction, and Ion Exchange Chromatography. J Liq Chromatogr Related Tech. 2009; Vol. 32 (15).
9. Valoran HP, Jeffrey SR. Ion Chromatography Analysis of Aminoglycoside Antibiotics. Applications of Ion Chromatography for Pharmaceutical and Biological Products. 2012: p. 175.
10. Jenke D. Application of Ion Chromatography in Pharmaceutical and Drug Analysis. J Chromatogr Sci. 2011; 49 (7): 524–39.
11. Karlsson E, Ryden L. Ion-Exchange Chromatography, in Protein Purification. Principles, High-Resolution Methods and Applications. 1998; Janson, J. C., Ryden, L, Ed., Wiley-VCH.
12. Hedhammar M, Karlström AE, Hober S. Chromatographic methods for protein purification. Royal Institute of Technology.
13. Hutin S, Brennich M, Maillot B, Round A. Online ion-exchange chromatography for small-angle X-ray scattering. Structural Biol. 2016; 72 (10): 1090 1099. doi:10.1107/S2059798316012833
14. Takano Y, Kashiyama Y, Ogawa NO, Chikaraishi Y, Ohkouchi N. Isolation and desalting with cation-exchange chromatography for compound-specific nitrogen isotope analysis of amino acids: application to biogeochemical samples. Rapid Commun Mass Spectrom. 2010 Aug 30;24(16):2317-23. doi: 10.1002/rcm.4651.
15. Ion Exchange Treatment of Drinking Water. Environmental fact sheet. New Hampshire Department of Environmental Sciences. 2009: WD-DWGB-2-12.
16. Stanton P. Ion-Exchange Chromatography. Methods Mol Biol. 2003; 25(12): 33-43. DOI: 10.1385/1-59259-742-4:23

CHAPTER 11

AFFINITY CHROMATOGRAPHY

INTRODUCTION

Biomolecules in living system interact specifically with other molecules in the cells. The basis for molecular recognition events lies in the affinities that form the complex machinery of all living beings. These affinities range from low- to very high-affinity interactions in the scales of picomoles. For a successful purification of a protein, each method is unique, and may require minor changes in processing, such as the upstream processes, and the methods applied to isolate a specific target protein. While deciding a suitable technique is important, a standard frame of reference can be applied to most proteins in combinations. The removal of particles from solutions and concentration of the proteins of interest will always be necessary, at the same time taking care to avoid any proteolytic degradation and denaturation.

Affinity interactions are used as a tool to study not only the cellular processes in normal and disease tissues, but also used to develop diagnostic and therapeutic products. In this chapter, the principles, techniques, variations, applications and future prospects of affinity chromatography will be discussed.

HISTORY OF AFFINITY CHROMATOGRAPHY

Emil Starkenstein, a German scientist reported the concept of 'resolving macromolecule complexes via their interactions with an immobilized'. The manuscript discussed the influence of chloride on the enzymatic activity of liveramylase and opened the door for developments in this field by several researchers (1). The term "affinity chromatography" was coined by Pedro Cuatecasas, Chris Anfinsen and Meir Wilchekin 1968, in their article that described the enzymatic purification technique using substrates and inhibitors that were immobilized (2).

This was followed by several reports that described Sepharose matrix activation using a cyanogen bromide (CNBr) reaction, as well as introducing a spacer arm to avoid any steric hindrance arising between the molecules that are intended to interact (2). Affinity chromatography has played an important role in many genomics, proteomics and metabolomic studies. The breakthrough in affinity chromatographic technique developments of affinity liquid has made studies such as biomolecule interactions, post translational modifications and protein degradation that were otherwise not possibly obvious.

PRINCIPLE

Affinity chromatography (AC) is a tool to separate proteins based on the principle of a reversible interaction between the target protein and a cognate ligand attached to the matrix of chromatography base. These interactions can be biologically specific, for example, a receptor binding a hormone; or biologically nonspecific, like a protein binding a specific dye substance.

In affinity chromatography, the binding of target protein to the ligand (the complementary binding substance) is specific and reversible. The conditions of binding are adjusted in such a way that favors sample binding to the ligand and the unbound material is subsequently washed out of the column. Following this, the bound target protein is eluted, or recovered by changing to suitable conditions, generally, by using a competitive ligand, or by changing pH, ionic strength, or polarity of the system. Ideally, the eluted target protein is concentrated and in in a purified form.

Through this technique, biomolecules can be separated through highly selective and specific interactions. This separation technique is based on the chemical structure or biological function of a given biomolecule. During the chromatographic process, the biomolecule's interacting partner (ligand) is immobilized on a chromatographic resin, also known as the stationary phase, while the desired biomolecule is present in the mobile phase, which is a mixture of many components. Finally, the material is eluted from the column by changing the composition of the mobile phase.

BIOMOLECULES PURIFIED BY AFFINITY CHROMATOGRAPHY

Antibodies were first purified using affinity chromatography isolate rabbit anti-bovine serum albumin antibodies. For their purification, bovine serum albumin was used as the affinity ligand on a cellulose support. Two years later, this technique was expanded to purify mushroom tyrosinase using an immobilized inhibitor of the enzyme (azophenol). Since then, affinity chromatography is commonly used to purify biomolecules such as enzymes, recombinant proteins, antibodies, and other biomolecules.

Affinity chromatography is often chosen to purify biomolecules due to its *excellent specificity, ease of operation, yield and throughput*. In addition, affinity chromatography has the ability to remove pathogens, which is necessary if the purified biomolecules are to be used in clinical applications. The purity and recovery of target biomolecules is controlled by the specificity and binding constant of the affinity ligand. In general, the association constants of affinity ligands used for biomolecule purification range from 103 – 108 M-1 (3). A common affinity ligand used in these purifications is an antibody, but other affinity ligands such as biomimetic dye-ligands, DNA, proteins and small peptides have been used as well. Figure 2 shows a wide variety of molecules that can be purified by affinity chromatography based on their polarity and volatility.

Table1: Some common interactions in affinity chromatography

Enzyme	Substrate analogue or inhibitor
Antibody	Antigen (virus, cells)

Nucleic acid	Complementary nucleic acid
Nucleic acid	Histone or other nucleic acid binding protein
Hormone	Hormone receptor
Glutathione	Glutathione S-transferase (GST) fusion protein
Metal chelate	His-tag fusion protein

COMPONENTS OF AFFINITY MEDIUM

For purification and separation of large biomolecules from complex mixtures using affinity chromatography, the ligand, the support (matrix), and spacer arms must be considered.

Ligand are the molecules that binds reversibly to a specific target molecule or group of target molecules

Matrix are used for ligand attachment and they should be chemically and physically inert.

Spacer arms are specifically used for the improvement of binding between ligand and target molecule by overcoming any effects of steric hindrance.

The matrix or affinity supports

The support materials for affinity chromatography are porous support materials such as silica, agarose, polymethacrylate, cellulose, and polyacrylamide, available as varied particle and pore sizes. While some supports are available with already immobilized commonly available affinity ligands (heparin, protein A), other types of support materials can be developed including flow-through beads, nonporous supports, membranes, supports that are monolithic, and expanded-bed adsorbents.

Nonporous support materials consist of nonporous beads with diameters of 1- 3 μm. While one of the major advantages this kind of support is that it allows faster purifications, these materials have low surface areas in comparison with traditional porous supports (4).

Membranes used in affinity chromatography also allow for fast separations like the nonporous beads, but lack diffusion pores and thus have limited surface area (5,6).

Initially, flow-through beads or perfusion media were developed for ion-exchange chromatography and possess both small and large pores. The large flow-through pores allow direct transportation of substances to the interior of the particle, thus requiring only short distances for diffusion (7,8).

For *monolithic supports*, the working principle is the same as perfusion media, and possess both large flow-through pores and small diffusion pores (9, 10).

To prevent column clogging *expanded-bed adsorbents* utilize the flow-reversal that allows column bed expansion and facilitates free flow of particulates through the column; this also prevents the column from fouling (11).

Several other factors, besides the type of support used in the affinity purification, must be considered when choosing a support material, including chemical -inertness, - stability, mechanical stability, particle and pore sizes.

Chemical inertness

The specificity of the binding depends on the affinity ligand immobilized onto the support, the affinity support should bind only the molecule of interest and have little or no nonspecific binding, thus, the need for chemically inert support material. Some supports which bind very little or do not exhibit any nonspecific binding, resemble the properties of the aqueous mobile phase. Due to this, support materials that are chemically inert are hydrophilic in nature. Since most separations are performed in media with low ionic strength, there should be minimum number of charges on the support to prevent interactions that are nonspecific.

Under optimal operating conditions, the support material should be chemically stable, which includes resistance to degradation by all enzymes and microbes, elution buffers, regenerating solvents, and cleaning agents that will be used within the column. These stability considerations must also be expanded to the stability of the affinity ligand matrix linkage. Agarose-based support materials meet all of these requirements as they can be used between pH 3 and 12, are not attacked by enzymes, and are not affected by most aqueous eluents. However, ligand attachment in agarose support materials is often not as stable, depending on the type of linkage used (12).

Mechanical stability

This is another consideration when choosing a chromatographic support material for affinity chromatography. Support materials must be able to withstand the backpressures encountered during normal separations without compressing. While most commercial packing materials meet this requirement, the build-up of particulate contaminants may restrict column flow and lead to high backpressures. Under these pressures, soft porous gel supports such as agarose beads will compress and increase the pressure even further causing collapse of the support structure. More mechanically stable supports (e.g. silica and heavily cross-linked polymers) are able to withstand these high pressures, but the build-up of particulate contaminants should be avoided if at all possible.

Particle size

The size of particles is an additional consideration when choosing a support material. Ideally, small particle sizes are desired to limit mass transfer effects and limit band broadening. In addition, smaller particle sizes tend to offer greater surface area of the support material and allow for a larger number of affinity ligands to be immobilized on the surface of the support. Unfortunately, as particle size is decreased, backpressures are increased. In addition, when using smaller particles, the potential for the build-up of particulate contaminants and column fouling is increased. For this reason, in preparative applications large particles (30 – 100 μm) are often used (13).

Pore size

Since the biomolecules of interest must be able to not only pass through the column but also be able to fully interact with the affinity ligand. Based on the estimation of the effective diffusion coefficient, the pore diameter should be at least 5 times the diameter of the biomolecule being purified (8). Therefore, a typical protein with a 60 Å diameter would need a support with at least a 300 Å.

LIGANDS USED IN AFFINITY CHROMATOGRAPHY

Antibodies

Antibodies have several advantages including their high specificity and relatively large binding constants. Antibodies or immunoglobulins are a type of glycoprotein produced when a body's immune system responds to a foreign agent or antigen. Due to the variability of the amino acid sequence in the antibody binding sites (Fab regions), it has been estimated that antibodies can be produced for millions or even billions of different foreign agents.

Antibodies which are produced by separate cell lines are referred to as polyclonal antibodies. Monoclonal antibodies are produced when a single antibody producing cell is combined with a carcinoma cell to create a hybridoma which can be grown in a cell culture. Monoclonal antibodies are often more desirable than polyclonal antibodies in affinity chromatography due to their lack of variability which allows for the creation of a more uniform affinity support.

Dye-ligand

Another type of affinity ligand which can be used to purify biomolecules from complex mixtures is a dye-ligand. Dye ligand chromatography originated in 1968 when Haeckel et al. were purifying pyruvate kinase using gel filtration chromatography and found that Blue Dextran (a small dye molecule) co-eluted with the protein. After further investigation, it was determined that binding between the dye and enzyme caused this coelution. The dye-enzyme binding was later utilized in the purification of pyruvate kinase using a Blue Dextran column in 1971 (14).

Biomimetic dye-ligands

Biomimetic dye-ligand chromatography takes dye-ligand chromatography one step further and utilizes modified dyes which mimic the natural receptor of the target protein. In addition to offering better binding affinities, these modified dyes were initially developed as a result of the concerns over purity, leakage, and toxicity of the original commercial dyes (Lowe et al, 1992). Cibacron Blue 3GA is one of the most common modified triazine dyes that has been used for protein purification.

Covalent attachment of the dye can be achieved through nucleophilic displacement of the dye's chlorine atom by hydroxyl groups on the support's surface (15).

Chlorotriazine poly-sulfonated aromatic molecules

Chlorotriazine poly-sulfonated aromatic molecules (triazine dyes) have been used for the purification of albumin, oxidoreductases, decarboxylases, glycolytic enzymes, nucleases, hydroloases, lyases, synthetases, and transferases (15). The main advantages of using dye-ligands and biomimetic dye-ligands are their low cost and resistance to chemical and biological degradation. The main disadvantage of these synthetic ligands is that the selection process for a particular biomolecule is empirical and requires extensive screening processes during method development. More information on biomimetic dyes can be found in reference (16).

DNA

DNA can also be used as an affinity ligand. It can be used to purify DNA-binding proteins, DNA repair proteins, primases, helicases, polymerases, and restriction enzymes. The scope of biomolecules which can be purified using DNA is expanded when aptamers are utilized. Aptamers are single-stranded oligonucleotides which have a high affinity for a target molecule. SELEX (Systematic Evolution of Ligands by Exponential Enrichment) allows for the isolation of these oligonucleotide sequences and allows for a wide range of potential targets including biomolecules which typically have no affinity for DNA or RNA.

RNA affinity ligands

A similar process can be used to develop RNA affinity ligands. Once a potential aptamer sequence is identified, it can be synthesized in vitro and used as the affinity ligand on a chromatographic support. An example of aptamers usage as in purification of L-selectin (17) and RNA binding proteins (18, 19).

Peptide affinity ligands

Peptide affinity chromatography is another method which can be used for purifying biomolecules. Peptide affinity ligands are typically identified using one of two techniques; biological combinatorial peptide libraries (e.g phage-displayed libraries) (20) or solid-phase combinatorial libraries (e.g. one-bead-one-peptide libraries) (21). Since then, peptide sequences have been isolated for a wide range of targets (22) and have been used to purify staphylococcal enterotoxin B, β-tryptase (23) and α-cobratoxin (24). The main advantages of using peptides as affinity ligands are their low cost and stability.

Other ligands can be used in affinity chromatography for biomolecule purification. For more information on all types of affinity ligands see references (25,26).

PURIFICATION METHOD

Purification by affinity starts with proper handling of the sample and the matrix, followed by selective binding (capture) of the target, washing to remove nonspecific background, and, finally, elution of the bound target. Successful affinity purification depends on a number of notable factors including, the amount and accessibility of the ligand on the resin, the strength of the interaction, and the integrity of protein to be immobilized. Usually, conditions are optimized to maximize the interaction between a target and immobilized ligand during the binding and wash process, then, switched to substantially weaken the interaction thus allowing for release of the target. It is recommended to perform small scale trials to select for the best purification conditions. A short summary of some common practical issues and considerations affecting affinity purification is given in the following sections.

Sample preparation: When preparing sample for purification, conditions should be selected to retain the proper fold and functionality of the target of interest. It is also highly recommended to remove insoluble materials and to reduce viscosity because both of these factors could clog the column, reduce the flow rate, and increase back pressure.

Some proteins tend to aggregate at high concentrations, which results in increased apparent molecular weight, decreased diffusion rate, and reduced capture by the affinity matrix. Dilution of the sample or cell lysate in a larger volume may be needed to reduce aggregation and increase protein capture and recovery under this circumstance. Conversely, if the sample is too dilute, binding rate and capture efficiency can be reduced especially for low-affinity binders.

Binding and wash

Efficiency of binding is related to the strength and the kinetics of protein–ligand interaction which can be affected by the nature of the interaction, the concentration of applied target, the amount of immobilized ligand, and the flow rate used for binding. The binding process can be simplified as in equation given below, assuming a 1:1 molar ratio where K_a is the association equilibrium constant, $[L]$ is ligand concentration, $[T]$ is the concentration of target protein, and $[LT]$ is the concentration of the complex. K_a equals $[LT]/([L] \times [T])$, which can also be expressed as k_a/k_d, where k_a is second order association rate constant that depends on the concentrations of both L and T, and k_d is the first order dissociation constant that does not depend on ligand concentration. Higher K_a usually leads to a higher adsorption ratio, defined as the ratio of bound to total applied target, thus, better binding. Normally, the ligand concentration on the matrix is around 10^{-2}–10^{-4} M and to achieve efficient binding, the K_a value should be in the range of 10^4–10^6 M^{-1}.

Affinity immobilization onto solid support can be achieved by passing the sample through a column packed with the affinity matrix, usually under ambient pressure and a slow flow rate. Generally, a higher flow rate will reduce the binding efficiency, especially, when the interaction between the ligand and protein is weak or the mass-transfer rate in the column is slow. The binding process can also be performed in batch, where the resin and sample are constantly mixed.

Batch binding

It promotes effective contact between target and immobilized ligand and often saves time, especially when dealing with large sample volumes; however, nonspecific binding can also increase. It is a good practice to optimize the amount of resin used during purification, where saturation of the resin with target during binding is preferred since excess resin can result in an increase in nonspecific binding as well as reduced target recovery due to readsorption, unless the latter is required under special circumstances.

Following binding, protein bound by nonspecific interactions can be removed by *washing*. For example, ionic interactions can be reduced by increasing salt (0.1–0.5 *M*) or changing pH values, and hydrophobic interactions can be removed by decreasing salt, altering pH, or adding surfactants (such as Triton X-100). Low amounts of competitive reagents can also be used to remove contaminants with weak affinity to the ligands. The flow rate and the volume (e.g., 5–10 column volumes) of the wash buffer should also be carefully determined for maximum removal of contaminants with minimum loss of target.

Elution

Elution of bound target from the resin is essentially the reverse process of binding, where conditions are optimized to reduce the K_a, that is, weakening the interaction between target and ligand. The elution condition should not denature the target protein, unless such conditions are compatible with downstream applications. There are two different types of elution methods:

Specific elution

The target protein–ligand complex is challenged by agents that will compete for either the ligand or the target thereby releasing the target protein into solution. The concentration and amount (volume) of competitive reagent used for elution will depend on their affinity relative to that of the immobilized complex, where weaker competitive reagents require higher concentration and more volume as compared to higher affinity additives. A good starting point for weak competitors is to use concentration 10-fold higher than that of the ligand. The specific elution is usually milder and proteins are more likely to retain their activity, but the slow elution, broad elution peaks, and the need to remove competing agent from the recovered protein are some of the drawbacks of this approach (25-28).

Nonspecific elution

For *nonspecific elution*, solvent conditions are manipulated to reduce the association rate constant, which ideally should approach zero, and to increase the dissociation rate constant, thus, weakening the overall affinity (K_a) resulting in dissociation of the complex.

Elution conditions can be optimized according to the mechanism of interaction between the ligand and protein, such as increasing salt concentration to reduce ionic interactions or by altering pH to change the protonation/ionization state, thus modulating the strength of hydrogen bonds, hydrophobic interactions as well electrostatic interactions. An example of elution by changing pH is the elution of antibodies from immobilized protein A or protein G, yet because the affinity of proteinA/G to antibodies is very strong, with K_a in the 10^8 M^{-1} range, a combination of different elution conditions may be required for maximum antibody release. In many cases, affinity requires proper three-dimensional folding of a protein so that chaotropic reagents or reagents that will affect protein folding can be used to elute target of interest; however, care must be taken to maintain proper folding of the target after elution by quickly returning to native conditions. When the affinity is weak, binding is achieved at high concentration of the target molecule which is then eluted by dissociation of the complex through dilution. This approach is known as isocratic elution (Cecilia et. Al.,2005).

$$L + T \underset{k_d}{\overset{k_a}{\rightleftarrows}} LT$$

THE KINETICS OF AFFINITY CHROMATOGRAPHY

The reaction between the ligand (L) and target compound (T) in an affinity atmosphere (either adsorption or desorption) is represented in Figure 1.

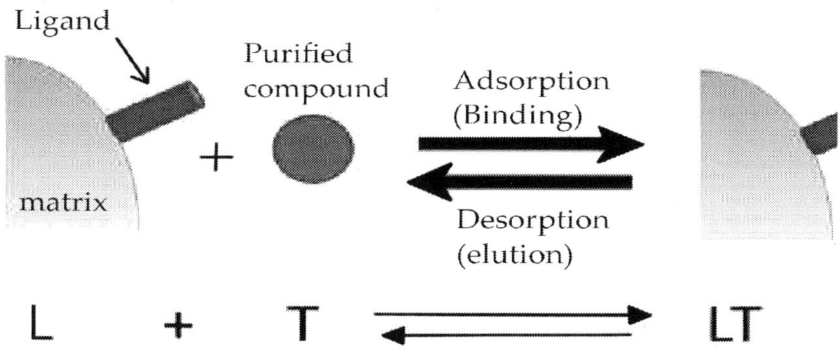

Figure 11.1. Basic reaction between compound to be purified and ligand.

The standard definition of the term equilibrium dissociation constant [K_D] can be expressed in equation 1,

$$K_D = \frac{[L]*[T]}{[LT]} \qquad (1)$$

where [L] is the free ligand, [T] is the target compound, and [LT] is the ligand-target complex.

According to the postulation of (Graves & Wu, 1974), the bound target-total target ratio can by represented as shown in equation 2,

$$\frac{\text{Bound target}}{\text{Total bound}} \approx \frac{L_0}{K_D} + L_0 \qquad (2)$$

where L_0 is the concentration of the ligand (usually $10^{-4} – 10^{-2}$ M). To achieve successful binding, the ratio of bound to total target must be near 1. Therefore, K_D should be small compared to ligand concentration. K_D can be greatly affected by changing in pH, ionic strength, and temperature. Thus changing these parameters can be used to control the binding and elution efficiency of the reaction and can be expressed as seen in equations 3 and 4,

$$L + T \rightleftharpoons LT \qquad (3)$$

$$LT \rightleftharpoons L + T \qquad (4)$$

In equation 3, the K_D range is between $10^{-6} – 10^{-4}$ M which means there is more binding and less elution. In equation 4, the K_D is decreased to $10^{-1} – 10^{-2}$ M by the elution conditions which results in less binding and more elution of the target compound.

The interactions described in equations 3 and 4 apply under non-selective (noncompetitive) elution conditions. In case of selective elution or competitive elution, the interaction can be represent as shown in Figure 11.2.

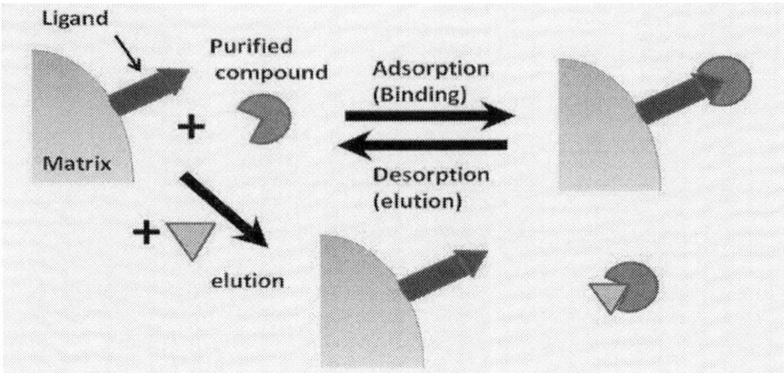

Figure 11.2. Competitive elution of the target by adding a competitive free ligand (triangle)

When adding a competing binding substance or a free ligand (C) that binds to the purified compound of interest during elution, the interaction can be represented as shown in Equation 5.

$$\mathrm{C} + \mathrm{T} \rightleftharpoons \mathrm{CT} \qquad (5)$$

The equilibrium constant, K_D, is calculated according to equation 6

$$K_D = \frac{[\mathrm{C}]*[\mathrm{T}]}{[\mathrm{CT}]} \qquad (6)$$

where [C] and [T] are the concentration of the free competing ligand and target, respectively and [CT] is the concentration of the competing ligand-target complex. (Graves & Wu, 1974) have shown that the eluted target to total target compound ratio can be represented by equation 7

$$\frac{\text{Eluted target}}{\text{Total bound target}} \approx \left({}^{p}\!/\!{}_{p+1} \right) * \left[\frac{pC_0}{pC_0 + \frac{K_{DComp} * L0}{K_D}} \right] \qquad (7)$$

where, p is the ratio between the volume of competitor added and the pore volume of the gel, K_D is the dissociation constant for coupled ligand, K_{Dcomp} is the dissociation constant for the free competing ligand, C_0 is the concentration of the competing ligand (usually $10^{-2} - 10^{-1}$ M) and L_0 is the concentration of the coupled ligand, usually $10^{-4} - 10^{-2}$ M. If both K_D and K_{Dcomp} are similar, then the concentrations of the competing and coupled ligand must be similar to achieve an efficient elution. On the other hand, if $K_{D\ comp}$ is equal to $5*K_D$ we would expect that the concentration of the competing ligand will need to be 5x higher to achieve successful elution.

APPLICATIONS OF AFFINITY CHROMATOGRAPHY

Immunoglobulin purification (antibody immobilization)

Antibodies can be immobilized by both covalent and adsorption methods. Random covalent immobilization methods generally link antibodies to the solid support via their free amine groups using cyanogen bromide, N-hydroxysuccinimide, N,N'-carbonyldiimidazole, tresyl chloride, or tosyl chloride. Alternatively, free amine groups can react with aldehyde or free epoxy groups on an activated support. As these are random immobilization methods, the antibody binding sites may be blocked due to improper orientation, multi-site attachment or steric hindrance.

Site-specific covalent immobilization

Site-specific covalent immobilization of antibodies can be achieved by converting the carbohydrate residues located in the Fc region of the antibody to produce aldehyde residues which can react with amine or hydrazide supports (29). Another site-specific immobilization of antibodies can be accomplished by utilizing the free sulfhydryl groups of Fab fragments. These groups can be used to couple the antibody fragments to an affinity support using a variety of established methods including epoxy, divinylsulfone, iodoacetyl, bromoacetyl, thiol, maleimide, TNB-thiol, tresyl chloride, or tosyl chloride methods.

Absorption onto secondary ligands

Antibodies can also be immobilized by adsorbing them onto secondary ligands. For example, if an antibody is reacted with hydrazide biotin, the hydrazide can react with oxidized carbohydrate residues on the Fc region of the antibody. The resultant biotinylated antibody can then be adsorbed onto an avidin or streptavidin affinity support. This type of biotin immobilization allows for site-specific immobilization of the antibody and can be performed using commercially available biotinylation kits.

Direct adsorption onto a protein A or protein G support

Alternatively, antibodies can be directly adsorbed onto a protein A or protein G support due to the specific interaction of antibodies with protein A and G. Immobilized antibodies on the protein A or G support can easily be replaced by using a strong eluent, regenerating the protein A/G, and re-applying fresh antibodies. Generally, this method is used when a high capacity/high activity support is needed. If a more permanent immobilization is desired, the adsorbed antibodies may be cross-linked to the support material using carbodiimide (30) or dimethyl pimelimidate (31,32).

Recombinant tagged proteins

Purification of proteins can be easier and simpler if the protein of interest is tagged with a known sequence commonly referred to as a tag. This tag can range from a short sequence of amino acids to entire domains or even whole proteins. Tags can act both as a marker for protein expression and to help facilitate protein purification.

The properties of fusion tags allow tagged proteins to be easily manipulated in the laboratory. Most significantly, the well-characterized tag-ligand chemistry enables single-step affinity purification of tagged molecules using immobilized versions of their corresponding affinity ligands. In addition, antibodies to fusion tags are also available and can be used for "universal" purification and detection of tagged proteins (i.e., without having to obtain or develop a probe for each specific recombinant protein).

In general, the most commonly used tags are glutathione-S-transferase (*GST*), histidine fusion (*His or polyHis tag*) and *protein A fusion tags*. Other types of fusion tags are also available including *maltose-binding protein* (33), *thioredoxin* (34), *NusA* (35), GB1 domain for protein G (36), and others (37,38). The decision to use any of these tagging methods depends mainly on the needs of the researcher.

GST tagged purification

Glutathione S-transferase (GST) is a 26 kDa protein (211 amino acids) located in cytosole or mitochondria and present both in eukaryotes and prokaryotes (e.g. *Schistosoma japonicum*). The enzymes have various sources both native and recombinantly expressed by fusion to the N-terminus of target proteins (39,40). GST-fusion proteins can also be produced in Escherichia coli as recombinant proteins. Separation and purifcation of GST-tagged proteins is possible since the GST tag is capable of binding its substrate, glutathione (tripeptide, Glu-Cys-Gly).

When glutathione is reduced (GSH), it can be immobilized onto a solid support through its sulfhydryl group. This property can be used to crosslink glutathione with agarose beads and, thus, can be used to capture pure GST or GST-tagged proteins via the enzyme-substrate binding reaction (Douglas, 1987). Binding is most efficient near neutral physiological conditions (pH 7.5) using Tris saline buffer and mild conditions to preserve the structure and enzymatic function of GST. As a result of the potential for permanent denaturation, denaturing elution conditions are not compatible with GST purification. In addition, upon denaturation or reduction, the structure of the GST fusion tag often degrades.

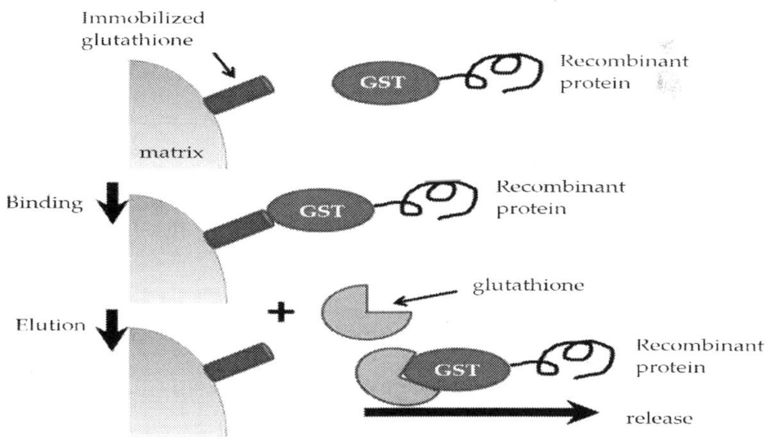

Figure 11.3. GST- tagged protein immobilization

Following a washing step to remove unbound samples, the bound GST-fusion protein can be recovered by the addition of excess reduced glutathione since the affinity of GST for free glutathione is higher than the affinity for immobilized glutathione. The free glutathione replaces the immobilized glutathione and releases the GST-tagged protein from the matrix allowing its elution from the column.

His-tagged protein purification

Recombinant proteins which have histidine tags can be purified using immobilized metal ion chromatography (IMAC). The His-tag can be placed on either the N- or C-terminus. Optimal binding and, therefore, purification efficiency is achieved when the His-tag is freely accessible to metal ion support (41).

Histidine tags have strong affinity for metal ions (e.g. Co2+, Ni2+, Cu2+, and Zn2+). One of the first support materials used immobilized iminodiacetic acid which can bind metal ions and allow for the coordination complex with the His-tagged protein. One difficulty with iminodiacetic acid supports is the potential for metal ion leaching leading to a decreased protein yield. Once the tagged protein is bound by the immobilized chelating agent, it can be eluted by introducing a competing agent for the chelating group (imidazole) or an additional metal chelating agent (EDTA).

One advantage of using His-tags for protein purification includes the small size of the affinity ligand. Due to the small size, it has minimal effects on the folding of the protein. In addition, if the His-tag is placed on the N-terminal end of the protein, it can easily be removed using an endoprotease. Another advantage of using His-tag purification methods is that *polyhistidine tags* can bind proteins under both native and denaturing conditions. The use of denaturing conditions becomes important when proteins are found in inclusion bodies and must be denatured so they can be solubilized.

Disadvantages of using His-tag protein purification include potential degradation of the Histag, dimer and tetramer formation, and coelution of other histidine-containing proteins. First, when a few histidine residues are proteolytically degraded, the affinity of the tagged protein is greatly reduced leading to a decrease in the protein yield. Second, once a protein has a His-tag added to its structure, it has the potential to form dimers and tetramers in the presence of metal ions. While this is often not a large problem, it can lead to inaccurate molecular mass estimates of the tagged protein. A third disadvantage of protein purification using His-tags is coelution of proteins that naturally have two or more adjacent histidine residues.

Lectin Affinity Chromatography

One particular technique that has seen growing interest over the last few years is lectin affinity chromatography. This is a type of affinity chromatography in which an immobilized lectin is used as the stationary phase. Lectins are non-immune system proteins that have the ability to recognize and bind certain types of carbohydrate residues (42). These carbohydrate-binding proteins are found in plants and animals, as well as in microorganisms (43). Due to their ability to bind to carbohydrate residues, lectins have been widely used to isolate and identify glycoproteins, glycopeptides, glycolipids and oligosaccharides (44). The most common types of lectins that are used in affinity chromatography are concanavalin A (Con A), wheat germ agglutinin (WGA), and jacalin (45). Con A specifically binds to targets that contain α-D-mannose or α-D-glucose residues. WGA binds to D-N-acetylglucosamine residues, and jacalin binds to galactose or mannose residues.

Lectins have often been used for sample pretreatment and target purification in applications where agarose is employed as the support. For example, in one recent study galactrox (i.e., a lectin that binds to galactose residues and that is found in snake venom from Botrox atrox) was purified using a lactosyl affinity column that made use of an agarose support (46). Small columns containing Sambucus nigra agglutinin (SNA) on agarose have been used for glycoprotein enrichment and to capture sialylated glycopeptides prior to glycosylation site mapping by mass spectrometry (47).

Sets of lectin columns have been employed for the isolation and purification of glycoproteins and glycoconjugates in an approach known as serial lectin affinity chromatography (S-LAC). This method has been used with small agarose columns containing Con A, WGA or SNA and a hydrophilic interaction column to extract and enrich glycoproteins and glycopeptides prior to their analysis by matrix-assisted laser desorption/ionization mass spectrometry (MALDI MS) and liquid chromatography/tandem mass spectrometry (LC-MS/MS) (48).

IMMOBILIZED METAL ION AFFINITY CHROMATOGRAPHY

Immobilized metal ion affinity chromatography (IMAC) is another type of affinity chromatography that has seen growing use in biomedical analysis. IMAC makes use of the specific interactions that can occur between immobilized metal ions and targets such as amino acids, peptides, proteins, and nucleic acids (49). The metal ions are immobilized within a column through the use of chelating agents like iminodiacetic acid, nitrilotriacetic acid, carboxymethylated-aspartic acid, and L-glutamic acid. Metal ions that are often chelated to these groups include Ni2+, Zn2+, Cu2+ and Fe3+ (50). IMAC was originally developed for the isolation and separation of metal- and histidine-containing proteins (51). During this type of analysis, the sample is passed through an IMAC column, and targets that can bind to the immobilized metal ions will be retained. These targets are later eluted through the addition of a competing agent (e.g., imidazole) and/or by changing the pH (52).

A growing application for IMAC has been in the detection of biomarkers for disease diagnosis. This type of work has often involved the use of IMAC with mass spectrometry in surface-enhanced laser desorption/ionization (SELDI) (53). The first step in this method is the immobilization of the desired metal ions onto a surface such as a chip. After excess metal ions have been washed away, the sample is applied and the target is allowed to bind to the immobilized metal ions while non-retained substances are washed away. An elution buffer is next added to release the retained target from the immobilized metal ions so the target can enter an analysis zone. An appropriate matrix is then placed on this surface to aid in the ionization and detection of the target by mass spectrometry. This technique has been used for detecting biomarkers in various samples, including serum, urine, and tissues.

ANALYTICAL AFFINITY CHROMATOGRAPHY

Besides its application for isolating and measuring specific targets, affinity chromatography can also be used to study interactions in biological systems. When affinity chromatography is utilized for this purpose, the method is referred to as analytical affinity chromatography, quantitative affinity chromatography or biointeraction affinity chromatography (54).

This approach has been used in many recent studies to examine drug binding with agents that include serum proteins, enzymes, and receptors, as well as the interactions of biomolecules with lectins, aptamers, and antibodies (55). Information that can be obtained by this approach includes the overall extent of binding, the equilibrium constants for an interaction, the number and types of sites involved in the binding process, and the rate constants for the interaction (57). One method that is often used to conduct these measurements is zonal elution. In this type of experiment, a small plug of a drug or analyte is injected onto an affinity column that contains the immobilized binding agent of interest. The retention factor or peak profile for the analyte is then analyzed under various conditions to obtain quantitative or qualitative information for a biological interaction.

Zonal elution

There are many different applications for zonal elution. For instance, retention factors can be used directly as a measure of the overall affinity of solute-ligand binding (58). This application has been used in HPAC for the high throughput analysis of drug interactions with HSA (59). The measurement of retention factors as the mobile phase or column conditions are changed (e.g., a variation in pH, ionic strength, solvent polarity or temperature) can also provide clues as to the forces and mechanisms that underlie a drug-protein interaction. Such an approach has been used in numerous studies to examine the chiral selectivity of proteins such as HSA, AGP, trypsin, α-chymotrypsin and cellobiohydrolase I (60). A comparison of retention factors that are measured by zonal elution for a set of structurally similar compounds can further be used to create quantitative structure-retention relationships that describe the interaction of these compounds with a given protein or binding agent, as has been used to examine drug binding sites on HSA and AGP (61).

Competition and displacement studies

Another common use of zonal elution experiments is in competition and displacement studies. In this type of experiment, injections of a target or probe are carried out in the presence of a fixed concentration of a competing agent in the mobile phase. Analysis of how the retention factor of the target changes with respect to the concentration of the competing agent can then be used to identify the type of competition that is occurring between these two solutes. In some cases, the same data can be further used to determine the association equilibrium constants and number of sites that are involved in the interaction.

Tool for studying the kinetics of a biological interaction

Another application of analytical affinity chromatography is as a tool for studying the kinetics of a biological interaction. There are several approaches for this type of work. One such approach is the plate height method, in which band-broadening is measured for a target on a column that contains the desired affinity ligand and on a control column that contains no ligand. The measured plate heights for the analyte are then plotted as a function of flow rate or linear velocity to find the plate height contribution due to stationary phase mass transfer, which is related to the dissociation rate constant between the target and affinity ligand (62). Two examples of this approach have been its use to examine the interaction kinetics of R/S-warfarin and D/L-tryptophan with HSA (63).

CONCLUSION

Affinity chromatography is a versatile technique that has become a valuable separation tool for biomedical and pharmaceutical analysis. This method can be carried out with traditional carbohydrate supports or with HPLC supports, such as silica and monolithic materials. This method can be used with step elution or isocratic elution and as part of affinity extraction or affinity depletion schemes. It can also be carried out with a wide range of ligands, such as lectins, boronates, antibodies, immobilized metal ions, molecularly imprinted polymers, dye ligands, and aptamers. These affinity methods can either be used alone or in combination with other techniques, such as reversed-phase HPLC or mass spectrometry. In addition, affinity chromatography can be used to study biological interactions and the competition or displacement of solutes on proteins or other binding agents. Based on the many recent applications of this method, it is expected that affinity chromatography will continue to grow in importance and popularity as a tool for the separation and analysis of biological and pharmaceutical agents in complex samples.

REFERENCES

1. Arsenis C, McCormick DB. Purification of flavin mononucleotide-dependent enzymes by column chromatography on flavin phosphate cellulose compounds. J Biol Chem. 1966, 241: 330-334.
2. Cuatrecasas P, Wilchek M, Anfinsen CB. Selective enzyme purification by affinity chromatography. Proc Natl Acad Sci USA. 1968, 61: 636-643.
3. Janson J-C. Large-scale affinity purification — state of the art and future prospects.
4. Trends in Biotechnology. 1984, 2: 31-38.
5. Wikstrom P, Larsson P. Affinity fibre: A new support for rapid enzyme purification by high-performance liquid affinity chromatography, J. Chromatogr. 1987;388: 123-134.
6. Charcosset, C. Purification of proteins by membrane chromatography. Jl. 1998;71: 95-110.
7. Roper DK, Lightfoot EN. Separation of biomolecules using adsorptive membranes. J. Chromatogr. A. 1995;702: 3-26.
8. Afeyan NB, Gordon NF, Mazsaroff I, et.al. Flow-through particles for the high-performance liquid chromatographic separation of biomolecules: perfusion chromatography. J Chromatogr.1990;519:1-29.
9. Gustavsson PE, Mosbach K, Nilsson, Larsson. Superporous agarose as an affinity chromatography support. J Chromatogr A. 1997;776: 197-203.
10. Mallik R, Hage DS. Affinity monolith chromatography. J Sep Sci. 2006;29: 1686-1704.
11. Tetala KKR, Van Beek TA. Bioaffinity chromatography on monolithic supports. J Sep Sci. 2010;33: 442-438.
12. Theodossiou I, Olander MA, Sondergaard, Thomas. New expanded bed adsorbents for the recovery of DNA. Biotechnol, Lett. 2000;22: 1929-1933.
13. Hermanson GT, Mallia AK, Smith PK. Immobilized Affinity Ligand techniques. Academic Press, San Diego,1992.
14. Clarke WS, Hage DS. Development of sandwich HPLC

microcolumns for analyte adsorption on the millisecond time scale. Anal Chem. 2001;73: 1366-1373.

15. Staal GE, Koster JF, Kamp H, van Milligen-Boersma L and Veeger C, Human erythrocyte
16. pyruvate kinase. Its purification and some properties. Biochim Biophys Acta. 1971;227:
17. 86-96.
18. Labrou NE, Clonis Y.D. Immobilized synthetic dyes in affinity chromatography, In
19. Theory and Practice of Biochromatography, Vijayalakshimi MA, 2002, 335-351, Taylor & Francis, London.
20. Clonis YD. Affinity chromatography matures as bioinformatic and combinatorial tools develop. J Chromatogr A. 2006;1101: 1-24.
21. Romig TS, Bell C, Drolet DW. Aptamer affinity chromatography: combinatorial chemistry applied to protein purification. J Chromatogr B Biomed Sci Appl. 1999;731: 275-284.
22. Dangerfield JA, Windbichler N, Salmons B, Gunzburg WH, Schroder R. Enhancement of the StreptoTag method for isolation of endogenously expressed proteins with complex RNA binding targets. Electrophoresis. 2006;27: 1874-1877.
23. Windbichler N, Schroeder R. Isolation of specific RNA-binding proteins using the streptomycin-binding RNA aptamer. Nat Protoc. 2006;1: 637-640.
24. Smith GP, Scott JK. Libraries of peptides and proteins displayed on filamentous phage. Methods Enzymol. 1993;217: 228-257.
25. Lam KS, Salmon SE, Hersh EM, Hruby VJ, Kazmierski WM, Knapp RJ. A new type of synthetic peptide library for identifying ligand-binding activity. Nature. 1991;354: 82-84.
26. Casey JL, Coley AM, Foley M. Phage display of peptides in ligand selection for use in affinity chromatography. Methods Mol Biol. 2008;421: 111-124.
27. Schaschke N, Gabrijelcic-Geiger D, Dominik A, Sommerhoff CP. Affinity chromatography of tryptases: design, synthesis and characterization of a novel matrix-bound bivalent inhibitor. Chembiochem. 2005;6: 95-103.

28. Byeon WH, Weisblum B. Affinity adsorbent based on combinatorial phage display peptides that bind alpha-cobratoxin, J Chromatogr B Analyt Technol Biomed Life Sci. 2004;805: 361-363.
29. Clonis YD. Affinity chromatography matures as bioinformatic and combinatorial tools develop. J Chromatogr A. 2006;1101: 1-24.
30. Hage DS. Handbook of Affinity Chromatography: Taylor & Francis, Boca Raton, FL, 2006.
31. Cecilia A, Roque A, Lowe CR. Advances and applications of de novo designed affinity ligands in proteomics. Biotechnol Adv. 2005;24: 17–26.
32. Graves DJ, Wu YT. On predicting the results of affinity procedures. Methods Enzymol. 1974;34: 140-163.
33. Ruhn PF, Garver S, Hage DS. Development of dihydrazide-activated silica supports for high performance affinity chromatography. J Chromatogr A. 1994;669: 9-19.
34. Phillips TM, Queen WD, More NS, Thompson AM. Protein A-coated glass beads. Universal support medium for high performance immunoaffinity chromatography. J Chromatogr. 1985;327: 213-219.
35. Schneider C, Newman RA, Sutherland DR, Asser U, Greaves MF. A one-step purification of membrane proteins using a high efficiency immunomatrix. J Biol Chem. 1982;257: 10766-10769.
36. Sisson TH, Castor CW. An improved method for immobilizing IgG antibodies on protein A-agarose,J Immunol Methods. 1990;127: 215-220.
37. di Guan C, Li P, Riggs PD, Inouye H. Vectors that facilitate the expression and purification of foreign peptides in Escherichia coli by fusion to maltose-binding protein. Gene. 1988;67: 21-30.
38. LaVallie ER, DiBlasio EA, Kovacic S, Grant KL, Schendel PF, McCoy JM. A thioredoxin gene fusion expression system that circumvents inclusion body formation in the E. coli cytoplasm. Biotechnology (N Y). 1993;11: 187-193.
39. Whetstone PA, Butlin NG, Corneillie TM, Meares CF. Element-coded affinity tags for peptides and proteins. Bioconjug Chem. 2004;15: 3-6.

40. Davis GD, Elisee C, Newham DM, Harrison RG. New fusion protein systems designed to give soluble expression in Escherichia coli. Biotechnol Bioeng. 1999;65: 382-388.
41. Balbas P. Understanding the art of producing protein and nonprotein molecules in Escherichia coli. Mol Biotechnol. 2001;19: 251-267.
42. Thorn KS, Naber N, Matuska M, Vale RD, Cooke R. A novel method of affinitypurifying proteins using a bis-arsenical fluorescein. Protein Sci. 2000;9: 213-217.
43. Allocati N, Federici L, Masulli M, Di Ilio C, Glutathione transferases in bacteria. FEBS J. 2009;276: 58-75.
44. Udomsinprasert R, Pongjaroenkit S, Wongsantichon J, Oakley AJ, Prapanthadara LA, Wilce MCet.al. Identification, characterization and structure of a new Delta class glutathione transferase isoenzyme. Biochem J. 2005;388: 763-771.
45. Dong XY, Feng XD, Sun Y. His-tagged protein purification by metal-chelate affinity extraction with nickel-chelate reverse micelles. Biotechnol Prog. 2010;26: 1088-1094.
46. Hage DS, Bian M, Burks R, Karle E, Ohnmacht C, Wa C. Bioaffinity chromatography. In: Hage, DS., editor. Handbook of Affinity Chromatography. 2nd ed. Boca Raton, FL: CRC Press; 2005. p. 107-135.
47. Liener IE, Sharon N, Goldstein IJ., editors. The Lectins: Properties, Functions and Applications in Biology and Medicine. Waltham, MS: Academic Press; 1986.
48. Yamamoto K, Tsuji T, Osawa T. Affinity chromatography of oligosaccharides and glycopeptides with immobilized lectins. In: Walker, J., editor. The Protein Protocols Handbook. Totowa, NJ: Humana Press; 2002. p. 917-931.
49. Bourne Y, Astoul CH, Zamboni V, Peumans WJ, Menu-Bouaouiche L, Van Damme EJM, Barre A, Rouge P. Structural basis for the unusual carbohydrate-binding specificity of jacalin towards galactose and mannose. J. Biochem. 2002; 364:173–180.
50. de Paula E, de Melo R, Sartim MA, Callejon DR, Paiva HH, Antonucci GA, et al. Isolation, functional, and partial biochemical characterization of galactrox, an acidic lectin from Bothrops atrox snake venom. Acta Biochim Biophys Sin. 2011; 43:181–192.

51. Vivekananda S, Nickens Z, Shah P, Sinnathamby G, Semmes OJ, Philip R. Investigation of sialylation aberration in N-linked glycopeptides by lectin and tandem labeling (LTL) quantitative proteomics. Anal Chem. 2010; 82:9201–9210.

52. Calvano CD, Zambonin CG, Jensen ON. Assessment of lectin and HILIC based enrichment protocols for characterization of serum glycoproteins by mass spectrometry. J Proteomics. 2008; 71:304–317.

53. Todorova D, Vijayalakshmi MA. Immobilized metal-ion affinity chromatography. Chap. 10. In: Hage, DS., editor. Handbook of Affinity Chromatography. 2nd ed. Boca Raton: CRC Press; 2005.

54. Li R, Wang Y, Guo LC, Shi M, Wang XG, Chen B, Zheng JB. A novel silica-based metal chelate stationary phase—L-glutamic acid-coppeR(II). Sep. Sci. Technol. 2011; 46:309–314.

55. Porath J, Carlsson J, Olsson I, Belfrage B. Metal chelate affinity chromatography: a new approach to protein fraction. Nature. 1975; 258:598–599.

56. Gutiérrez R, Martín del Valle EM, Galán MA. Immobilized metal-ion affinity chromatography: status and trends. Sep Purif Rev. 2007; 36:71–111.

57. Felix K, Fakelman F, Hartmann D, Giese NA, Gaida MM, Schnoelzer M, Flad T, Buechler MW, Werner J. Identification of serum proteins involved in pancreatic cancer cachexia. Life Sci. 2011; 88:218–225.

58. Chaiken, IM., editor. Analytical Affinity Chromatography. Boca Raton, FL: CRC Press; 1987.

59. Patel S, Wainer IW, Lough WJ. Chromatographic studies of molecular recognition and solute binding to enzymes and plasma proteins. In: Hage, DS., editor. Handbook of Affinity Chromatography. 2nd ed. Boca Raton: CRC Press; 2006. p. 663-683.

60. Schiel JE, Hage DS. Kinetic studies of biological interactions by affinity chromatography. J Sep Sci. 2009; 32:1507–1522

61. Schiel JE, Joseph KS, Hage DS. Biointeraction affinity chromatography, Chapter 4. In: Grinsberg, N.; Grushka, E., editors. Advances in Chromatography. Vol. Vol. 48. New York: Taylor & Francis; 2009.

62. Hage DS, Chen J. Quantitative affinity chromatography: practical aspects. In: Hage, DS., editor. Handbook of Affinity Chromatography. 2nd ed. Boca Raton: CRC Press; 2006. p. 595-628.

63. Kim HS, Wainer IW. Rapid analysis of the interactions between drugs and human serum albumin (HSA) using high-performance affinity chromatography (HPAC). J Chromatogr. B. 2008; 870:22–26.

64. Xuan H, Hage DS. Evaluation of a hydrazide-linked alpha1-acid glycoprotein chiral stationary phase: separation of R- and S-propranolol. J Sep Sci. 2006; 29:1412–1422.

65. Kalizan R, Noctor TAG, Wainer IW. Stereochemical aspects of benzodiazepine binding to human serum albumin. II. Quantitative relationships between structure and enantioselective retention in high performance liquid affinity chromatography. Mol Pharmacol. 1992; 42:512–517.

66. Schiel JE, Hage DS. Kinetic studies of biological interactions by affinity chromatography. J Sep Sci. 2009; 32:1507–1522.

67. Loun B, Hage DS. Chiral separation mechanisms in protein based HPLC columns 2. Kinetic studies of (R) - and (S) -warfarin binding to immobilized human serum albumin. Anal Chem. 1996; 68:1218–1225.

CHAPTER 12

SIZE EXCLUSION CHROMATOGRAPHY

INTRODUCTION

Size-exclusion chromatography (SEC) is a chromatographic method in which molecules in a sample are separated by their size or molecular weight. Size-exclusion chromatography (SEC) and gel-permeation chromatography (GPC) are two names are based on the same principle with the former accounting for separation of biomolecules based on size and the latter is based on separation of polymers (1). The technique has been favoured for application in an evaluation of aggregation of proteins that is a vital study especially in the pharmacy industry (2).

HISTORY

The potential of separating molecules based on size using chromatography found its origin in the theories in 1950 (3) from previous studies on zeolites whose tiny pores aided in the exclusion of small molecules (4). The separation based on size was applied in the work published in 1953 where ionic resins where applied to separate nonionic molecules by Wheaton and Bauman (5).

A subsequent study examined the use of cation exchange resin for the separation of glycerol from polyhydric alcohols (6). The exclusion based on size was performed by Lindqvist and Storgårds where a starch based column was used to separate peptides (7). The technique of using a molecular sieve effect was applied using starch as a material for columns. As starch has low affinity of proteins several proteins were separated using such columns (8).

PRINCIPLES BEHIND SIZE EXCLUSION CHROMATOGRAPHY

The free energy change of a chromatographic process can be described by

$$G^0 = \Delta H^0 - T\Delta S^0 = RT\text{In}\,k \qquad (1)$$

where ΔG^0, ΔH^0, and ΔS^0 are the differences in standard free energy, enthalpy, and entropy, respectively; R is the gas constant: T is absolute temperature, and k is the partition coefficient.

For most chromatographic modes of separation, the enthalpy of adsorption contributes a major amount to the net change in free energy. SEC stands apart in the principle of partitioning is driven entirely by entropic processes with zero absorption in an ideal condition which makes $\Delta H = 0$. Hence, the previous equation can be modified as:

$$\text{In}\,K_D = -\Delta S^0 / R \qquad (2)$$

K_D is the thermodynamic retention factor in SEC.

Thus, in SEC separations, temperature should have no impact on retention. In practice, temperature can indirectly impact retention to a small degree by altering the conformation of the proteins, as well as by affecting the viscosity of mobile phase and diffusion properties of analytes.

The thermodynamic SEC retention factor is defined as the fraction of intra particle pore volume that is available to an analyte:

$$K_D = \frac{V_R - V_0}{V_i} \qquad (3)$$

where V_R, V_0, and V_i are the retention volumes of the analyte of interest, the interstitial volume, and the intra-particle volume respectively.

K_D can have a value of 0 where the analyte is fully excluded from the pores of the stationary phase, upto unity where the analyte can entirely access the intraparticle pores. Rearrangement of Eq. (3), yields:

$$V_R = K_D.V_i + V_0 \qquad (4)$$

Ideally the separation of molecules by SEC is based on the Stokes radii of the analytes in solution.

The size based separation involves calibration curves are based on proteins or polymers of known molecular weight. This calibration curve: log M vs. the retention volume yields a third order polynomial, with a linear region that can be used to estimate the molecular weight of a sample or unknown analyte.

The pore size of the matrix used for packing the column shows a high influence on the calibration curve that is used to estimate the range of molecular weight. Depending on the Stokes radius of the analytes, the size of the pores and their configuration allows or restricts them. The first to be eluted are the large sized molecules as they are excluded from the pores followed by the samples of descending sizes to be eluted.

Macromolecules such as proteins show several patterns such as fibrous or globular posing a challenge in the deduction of molecular weight. An additional challenge is variation in the absorption than what is expected in an ideal prediction. Such challenges warrant the use of universal calibrants.

The slope of the line in the linear portion of the calibration curve reflects the selectivity of the stationary phase, expressed by the following relationship:

$$\log M = m.K_D + b \qquad (5)$$

where m and b are the slope and intercept of the line respectively.

With reduction in the pore sizes the slope reduces to enable a finer discernment of molecules that have a similar size. The point at which K_D = 0.5 is a measure of the mean pore size of a particle.

A term k" was introduced as the ratio of mass of solute inside the particle to that outside the particle expressed as follows: [11]

$$k'' = K_D \frac{V_i}{V_0} \qquad (6)$$

When this is introduced in (5), the decrease in V_0 or an increase of V_i *can be used to fine tune the selectivity.*

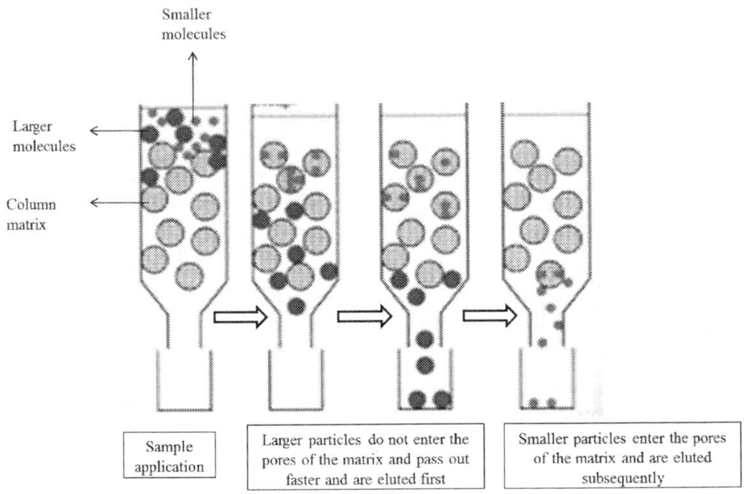

Figure 12.1: Principle of size exclusion chromatography

KINETICS AND RESOLUTION

The Van Deemter Equation is expressed by the following relationship:

$$H = ad_p + bD_m / u + cud_p^2 / D_m \approx ad_p + cud_p^2 / D_m \qquad (7)$$

Where plate height is H when linear velocity u, diffusion coefficient D_m, and particle size d_p,

The plate height and particle size are dependent as shown by this equation that implies that a reduction of the particle size can enhance the efficiency of an experiment.

A relationship has been used for the calculation of diffusion coefficients of proteins

$$D_m = 8.34.10^{-10}.\frac{T}{\eta M^{1/3}} \tag{8}$$

where T is the absolute temperature in Kelvin, is the mobile phase viscosity in Poise, and M is the molecular weight of the protein (12)

Resolution

The term resolution in chromatography, is expressed as:

$$R_s = \frac{V_{R2} - V_{R1}}{4\sigma} = \frac{\Delta V_R}{4\sigma} \tag{9}$$

where the subscripts 1 and 2 represent the respective analyte, and σ is the mean standard deviation of the respective peak widths.

The slope m in the linear region of the calibration curve as defined in previously in the equations (3) and (5):

$$m = V_p \frac{\log M_1 / M_2}{\Delta V_R} \tag{10}$$

A combination of the following expression (11) and Equation (7)

$$N = \frac{V_R^2}{\sigma^2} = \frac{L}{H} \tag{11}$$

Results in the following equation for resolution in the chromatography experiments that includes the column length, pore volume and linear velocity [2].

$$R_s = -\frac{1}{4}.\frac{V_p}{V_R}.\frac{\Delta \log M}{m}.\sqrt{\frac{L}{ad_p + cud_p^2 / D_m}} \tag{12}$$

MATERIALS USED AS PACKING MATERIALS IN A COLUMN

As discussed initially, the initial columns were made of starch that offered certain challenges such as its lack of strength or its poor characterization due to its natural origin. This led to the development of dextran molecules that were linked with epichlorohydrin that offered advantages such as very less interaction with analytes such as proteins and enhanced mechanical strength. This *dextran* was commercialized by Pharmacia that gave it the brand name, Sephadex: the size of the pore network of the dextran depends on the degree of crosslinking, that aids in the determination of the range of biomolecules that can be separated (9, 10,12).

Initially, *Sephadex* was manufactured in the form of irregular particles and later synthesized as porous spheres. Subsequent variation in the level of crosslinking could enable the retaining or exclusion of analytes.

The other materials involved include *agar, agarose, polyvinylethylcarbitol or polyvinylpyrrolidone and also polyacrylamide.* These polyacrylamide gels were commercialized by Bio-Rad under the trade name Bio-Gel. Subsequent years saw the use of porous silica and its derivatives for use as columns, as it presented several advantages such as its ability to not swell, better mechanical strength and its non-reactive nature. The molecules called *silanols* on the surface of the silica interacted with protein samples that posed a problem. This warranted the use of modifications of surfaces using additives: such as glyceropropylsilane and acetylaminopropylsilane. These additives show reactions with protein samples using hydrophobic reactions that turned research towards the use of diol based groups as modifiers as they circumvent the issue of hydrophobic interactions (10).

Hydrophilic polymers were separated using silica while hydrophobic polymers were separated using silanes. For materials that are insoluble in water, the use of cross-linked styrene finds widest use. A 2003 work reported the use of ethyl bridged molecules of organic as well as inorganic origin (Wyndham et al, 2003). Further work saw the use of particles with their surfaces modified by trimethylsilyl for the separation of organic molecules. These molecules were reported to possess lower acidity and lesser unwanted interactions with analytes (1).

Hydrophobic polymers saw the use of *porous gels* by the Dow Chemical Company. The column involved the preparation of divinylbenzene that were crosslinked. The gels involved the loading of the material of dimensions 44–75-μm into a tube to separate molecules within satisfactory time frames (12).

Bridged ethyl hybrid (BEH) *particles* with diol modifications on their surfaces have also shown a reduction in the acidity apart from the greater strength and a reduction in the size of particles to ultimately increase the efficiency of the separation.

Apart from porous particles that find use in SEC columns, there are reports of sorbents as packing materials. A porosity of about 35–41% has been reported in the interstitial spaces that does not offer any significant advantage in separation and also increases the time for analysis (2).

A 1990 work reported the use of aligned porous fibers arranged in bundles that caused a decrease of the interstitial porosity to 15–18%. This resulted in an increased pore volume between the particles (13). Another study in 2009 studied the use of monoliths composed of poly (ethylene glycol methyl ether acrylate-co-polyethylene glycol diacrylate). They showed similar results to a packed bed of ca. 8 μm particles especially in terms of lesser amount of protein binding and efficiency. These columns could resolve peptides across a vast range of molecular weights of 670,000 Da and less (14).

PROTOCOL: A GLIMPSE

The separation of molecules using SEC involves packing of a medium in the form of a packed bed following which it is subjected to equilibration with buffer. As the buffer occupies the space between particles, equilibrium is reached between the particles of the stationary phase and that of the mobile phase. This is followed by elution of the samples, in an isocratic fashion without the use of several buffers. A running buffer can be applied to wash the column as a last step in the protocol to eliminate certain groups that remain and the column can be subjected to another cycle. A major *advantage* is that SEC can be used as an immediate step following other chromatographic separations such as affinity chromatography and ion exchange chromatography.

Following application of a sample, depending on the volume of the packed bed in the column, the experiment is carried out using one column volume of a single buffer (isocratic elution). In order to achieve maximum resolution in a rapid fashion, the parameters of flow rate and column dimensions come into play. Each experiment requires optimization to achieve balance between these two parameters. It has been observed that a long column and a low flow rate yields high resolution while short column and a high flow rate yields the most rapid outcomes.

Modes:

SEC can be applied in *group separation mode* to eliminate small molecules from a group of larger molecules and as a rapid and elementary solution for buffer exchange. Small molecules such as excess salt or free labels are easily separated from larger molecules. Samples can be prepared for storage or for other chromatography techniques and assays. Protein purification protocols employ SEC in group separation mode especially for desalting and buffer exchange. Materials commonly reported include Sephadex G-10, G-25, and G-50. The use of wide and small columns can facilitate tha application of large sample volumes, 30% of the total column volume (packed bed).

SEC can also be used in *fractionation mode* to separate several components in a mixture on the premise of differences in their size. The method can result in the isolation of one or more of the components, or the evaluation of the molecular-weight distribution in the sample.

This process finds application as High-resolution fractionation that is applicable as a final polishing step in a purification scheme. The separation of monomers from aggregates has been reported with ease. Subsequently, samples can be transferred to an appropriate buffer for assay or storage.

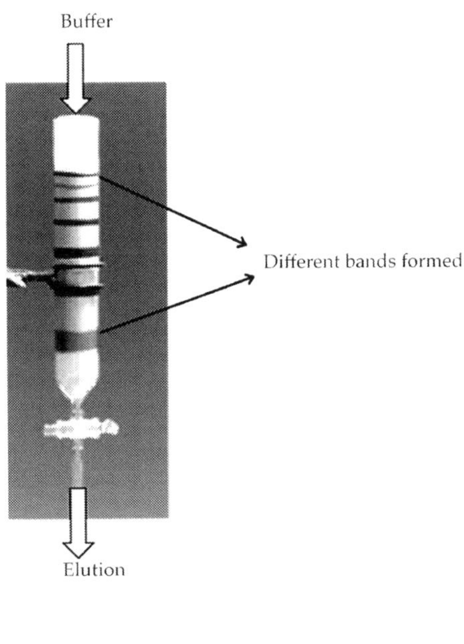

Figure 12.2: Scheme of a size exclusion experiment

Volume considerations:

Sample volumes are required to be ascertained through trial and error. 30% of the total column volume can be applied for group separations while 0.5% to 4% of the total column volume can be applied in high-resolution fractionation. The point of concern here is that sample volumes should be less than 2% in order to facilitate good resolution. Though, more sample volumes can be applied on a column, it would depend on the samples and the experiment in question.

The column volume depends on the sample volume that would be applied. It is recommended that a sample volume that is 0.5% of the total column volume can be applied for the separation of a sample or for analysis. The use of a large sample volume would necessitate the use of a large column volume that can involve the repetition of the process multiple times in a small column, followed by pooling and concentration of the samples.

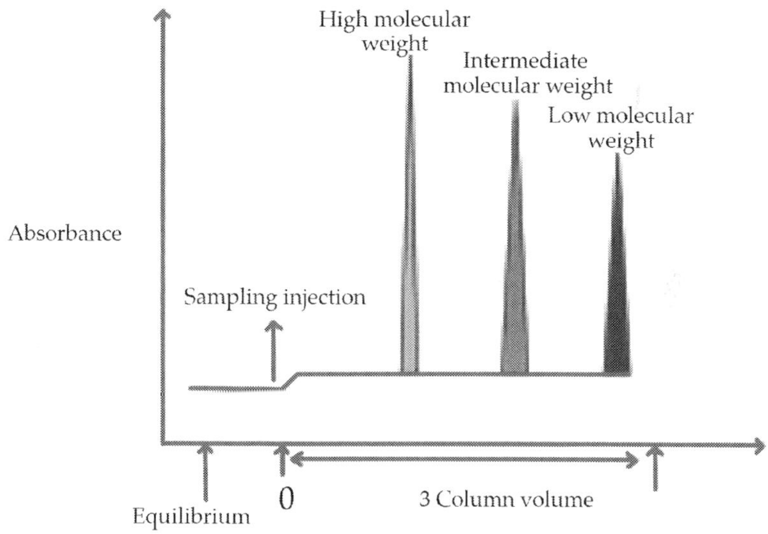

Figure 12.3: A chromatogram: Graphical description of separation: Following application of a sample, the smallest molecule (blue) is more delayed than the largest molecule (yellow); the latter is eluted first.

Packing: Group separation experiments can use columns that have the matrix packed up to dimensions of 100 mm while a packing of 300 to 600mm is suitable for high-resolution fractionation. A height of 150mm can be used for quick screening protocols. The packing is a vital point in the separation of molecules or purification as more narrow peaks are displayed in the output when the packing is efficient. The efficiency of the column must be determined in a routine manner to ensure smooth and consistent operations. The checking can be carried out using the evaluation of peak symmetry as well as the plate numbers taking into consideration factors such as dead volume and capillaries used (15).

INSTRUMENTATION

The initial years of SEC saw the separation of molecules such as proteins on SEC using the native conditions in order to enable the molecule in study to maintain its activity. These native conditions entail the use of certain conditions such as elevated levels of salt, maintenance of body pH in the experiment as well as mobile phases that are largely aqueous. It also requires the use of systems that are biocompatible in order to avoid or decrease the formation of unwanted interactions of the analytes and also the formation of complexes. However, the increased levels of salt have been reported to adversely compromise the efficiency of the apparatus and the column. The use of these aqueous mobile phases can result in the growth of microbes that warrants the use of certain anti-microbial compounds in the setup.

A major focus in the use of SEC includes the volumes of the recorded peaks. Unlike adsorption techniques where larger volumes are applied in the system than the columns used to achieve separation, SEC uses only a column volume to achieve the separation. The retention volume plays a major role in the determination of the peaks as the width of a peak increases when the peak traverses the column. The use of columns with dimensions especially a radius of 3.75mm or more is warranted in SEC in order to lessen the effect of pressure that can cause dispersion and in turn cause peaks to widen (2).

This requires optimization of the systems used in SEC for high performance. A common practise is to decrease the tubes involved in the system and also use tubes with very minute dimensions such as not more than 0.0025” in radii. It has been reported that the use of increased volume of the system especially through the use of guard columns or detectors or injecting systems can adversely affect the efficiency of an experiment in SEC (16).

The use of detectors that use the principle of light scattering can cause an increase in the band dimensions. Again, such an effect depends on the samples used especially in terms of its dispersion status, with samples that have low dispersions not getting much affected. Such samples show very little shifts in volumes in SEC and yield results that are well in the required range of molar masses reported (17).

Resolution of SEC has been improved with the development of columns that use reverse phase techniques. Initially the matrix had dimensions of 2 micron that required the use of instruments with low dispersion that have been extended to the use of sub 2 μm particles. Such instruments have enhanced the resolving power of experiments conducted in SEC (18).

DETECTION OF SAMPLES

In experiments that involve the use of purification studies of proteins, the samples require the analysis of the samples. The samples can be subjected to several analyses such as mass spectrometry, immunological techniques or analysis of the spectrum of the target molecule. A major technique to assess the purity is SDS-polyacrylamide gel electrophoresis (SDS-PAGE) that also can be run with a range of standards of known molecular masses to ascertain or confirm the mass of the sample under study. Few techniques that have been reported include:

Colour based techniques such as Bradford total protein assay or Lowry.

The quantification of yield and purity can be checked by evaluating the amount of absorbance at 280 nm.

Immunological techniques such as immunoprecipitation or ELISA or western blots can be carried out in order to check the samples.

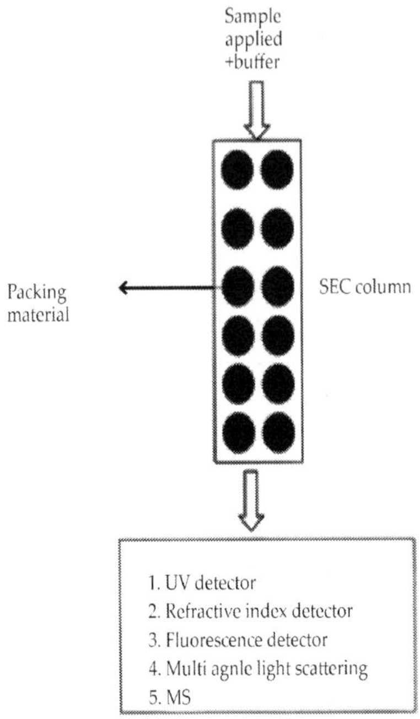

Figure 12.4: An overview of an experiment with several detectors.

A summary of several techniques used in detection of samples are given below:

UVWavelength:

The measurement of proteins can be achieved by using UV wavelength that can target aromatic amino acids with more preference. The use of 214 or 220 nm wavelengths can target the detection of amide bonds. Either of the wavelengths used has several advantages such as, a low wavelength, that results in a more sensitive detection while a larger spectrum of samples can be attained by larger wavelengths (2).

Dual wavelength detection:

This technique uses an amalgamation of detection at both low and high wavelengths to maximise the individual merits. The ratio of the wavelengths are determined for the samples to be detected. The smaller wavelength can detect impurities or aggregates formed and makes the detection more sensitive while the larger wavelength yields a broader range of samples.

Such a dual wavelength detection has been successfully applied in the detection of aggregates in an antibody sample that led to better characterization of its degradation pathways (19).

Fluorescence detectors:

These detectors have resulted in an improvement of the technique especially in terms of sensitive detection. A 2010 study evaluated the use of fluorescence detection of Interferon Alpha-2 in an SEC based study. The team reported an increase in the sensitivity of the technique in the analysis of the protein of choice that shares close elution profile with excipients added to pharmaceutical preparations, through the use of fluorescence based techniques (20).

Multi-angle light scattering (MALS):

These detectors use the multi-angle light scattering in order to study the shape or size of proteins. This aids in the combination of SEC based experiments with that of SEC to achieve separation as well as data on molecular weight. The combination of MALS and SEC has aided a better analysis in the analyzing of molecular weights and also yield data on the concentration, shape and size of the analyte.

The detectors require data on extinction coefficient and concentration of analytes that requires the use of an additional system for the measurement of concentration. This can use ultra-violet (UV) or refractive index (RI). A 2003 study evaluated the efficiency of SEC-MALS in the analysis of Aggregates of Human Ig G. It was observed that detection using UV could not yield highly sensitive detection of aggregates while MALS could detect the aggregates especially in trace amounts of such complexes (21).

Mass spectrometry/ MS:

MS is another technique that yields data on molecular weight. However, certain challenges have been reported in the coupling of SEC with MS. One of the obstacles is the usage of a mobile phase that has large amounts of salts that are not volatile. Such phases cause suppression of ions in addition to contamination of the instrumentation (22).

This challenge can be surmounted by the use of a mobile phase based on organic compounds. For instance, organic solvents were applied in the purification of immunoconjugates with a high sensitivity reported in the detection of conjugates against antibody preparations (23).

FACTORS AFFECTING SEC

Interactions:

Ionic interactions between proteins and the particles used to pack columns are a major factor as they can result in the alterations of the proteins or the quality of the peaks. This requires the use of surface modifiers that has been discussed in the materials used.

In addition to ionic interactions, electrostatic interactions can also play a role in the experiment. For instance, opposite charges between analytes and column matrix can cause an adsorption of the analyte that results in an increased time of elution. A decrease in such electrostatic interactions can be achieved by an increase in the concentration of salt to elevate the ionic strength of the medium. A 2004 study showed that of several concentrations of sodium chloride added in the SEC of Synthetic hPTH (1–34), a concentration of 100 mM caused a decrease in such interactions and enhanced the dispersion of aggregates (24).

Another approach to minimize electrostatic interactions involves the addition of a counter ion in the mobile phase. This approach has a limitation as such counter charged species can cause hydrophobic interactions between proteins. Hydrophobic interactions that involve the association of an analyte with hydrophobic portions of the packing material that causes an impediment to quick elution.

Secondary interactions can be checked by the use of certain molecules in the mobile phase especially in the realm of SEC where the level of aggregates is required to be low. Certain studies involving SEC of therapeutic Proteins have successfully reported the use of arginine to the mobile phase. It has been reported that the amino acid binds the analyte in question and prevents its interaction with packing materials. It was also observed that the addition of arginine caused an enhanced recovery of samples from aggregate formation (25). Another study has reported that when arginine was added to the mobile phase, the peaks were found to be more perfect as compared to controls (26).

These interactions can be controlled by looking at the pH of the mobile phase used in SEC. Based on the isoelectric point of the protein and the pH of the mobile phase, the conformation of an analyte can be subject to change which in turn affects its interaction with the packing material. An early study observed that, while ion-exclusion was reported at pH values above the isoelectric point of the protein ion exchange was reported at pH values below the pI of the protein.

Column Dimensions:

Looking at the Resolution equation previously mentioned:

$$R_s = \frac{1}{4} \cdot \frac{V_p}{V_R} \cdot \frac{\Delta \log M}{m} \cdot \sqrt{\frac{L}{ad_p + cud_p^2 / D_m}} \tag{12}$$

The resolution shows proportionality with the square root of the column length. This implies that when the column length is increased, resolution can be boosted (2)

The length of columns can be increased by the attachment of multiple columns in links. For instance, a 1996 study linked 2 columns for the analysis of proteins: ovalbumin and BSA. Though, the efficiency of the column and resolution were enhanced, the run time also showed an increase (27).

Flow Rate

Another key factor in the design of a SEC experiment is the Flow rate: with low velocity in a linear fashion yielding a good run. The same study listed above observed that reduction in the flow rate caused a boost in the resolution achieved with SEC of ovalbumin and BSA (27).

Sample Load

The resolving power of a SEC run can be decreased if a column has an excess of sample. The 1996 study reported above also examined the adverse effects of overloading on the peaks of SEC: When a range of samples were applied on the system from 2–200 μL: it was observed that when the sample volume crossed 100 μL, the peaks were broadened and resolution was diminished between the protein samples tested (27).

An ideal and optimum range reported is sample volumes between 5–10% of the total column volume, beyond which the resolution of the system is affected (2).

HIGH POINT OF SEC

SEC has the advantages of a quick run of the experiment apart from the technique yielding highly reproducible results. The development of SEC columns that use sub-2 μm packing materials used in conjunction with low dispersion instrumentation has caused a decrease in the system run time. HPLC columns have a run time of 15–20min, while the SEC system that uses these micrometer particles report sample interlacing routine analyses of less than 120s per

CHALLENGES WITH SEC

The technique of SEC requires validation of its results especially in terms of reports of the amount and status of aggregates in a sample. This validation can be carried out by additional methods such as dynamic light scattering, asymmetrical flow field flow fractionation or sedimentation velocity analytical ultracentrifugation when the samples to be checked are products for use in therapy.

In addition to evaluating the levels of aggregates, additional studies include the modifications or the activity or the details on mass. Such studies require repeated validation especially as an analyte in question traverses from an initial lab stage to a final stage (2).

APPLICATIONS

There have been numerous applications of SEC reported for several analytes. The supervision of proteins or conjugates such as protein-drug conjugates and their structures has been done by SEC (29).

Biotherapeutics:

SEC has been reported to be of highly sensitive and reproducible to be used for the evaluation of protein aggregation (30). The European Pharmacopoeia guidance document that has the title, entitled "Technical Guide for the Elaboration of Monographs on Synthetic Peptides and Recombinant DNA Proteins," enlists SEC in the analysis of aggregation of proteins. Additionally, SEC finds its place in a major portion of Chemistry, Manufacturing & Controls (CMC) sections as a test in regulatory filings.

The analysis and quantitation of dimers, trimers, and higher order aggregates for a wide variety of biologic-based therapies, such as insulin, recombinant human growth hormone (rGH), and monoclonal antibodies is vital to avoid safety issues. Such analysis of aggregates use various techniques that are biotherapies based on the size of the biomolecules (31).

Several techniques asymmetric flow field flow fractionation (AF4), SV-AUC, multi-angle light scattering (MALS), and SEC are employed, the dominant method continues to be SEC (32).

The technique has been applied mainly in the field of biotherapeutics for evaluating the levels of aggregates that lead to the formation of high molecular weight forms that can impact the efficiency and the safety of products. This is of vital importance to products such as monoclonal antibody (mAb) preparations whose status or amount of soluble protein aggregation forms part of its critical quality attributes.

On one end, there are low valency high molecular forms of a soluble protein, that can serve as a window to examine the process and product. Such aggregations can be linked to the actual process or the culture conditions or the last stages of formulation. On the other end, the study of high valency high molecular forms is essential with reports of such complexes that stimulate an immune response (33). SEC is sensitive, reproducible and offers a high sample throughput to find application in the analysis and the measurement of such high molecular forms in biotherapeutics. However, non- specific binding to the columns or filtration can yield an imprecise reflection of the amount of aggregates. This warrants a further corroboration with other techniques such as dynamic light scattering or sedimentation velocity analytical ultracentrifugation (1).

SEC and formulation studies:

The technique can be applied in the formulation and manufacturing techniques. The selection of cell lines can be done using SEC to narrow down on a cell line that synthesizes products with minimum aggregates. An added application is the analysis of the various forms of aggregates that are formed of a product in order to get an idea of planning the downstream purification (Jing, 2012). The technique can also aid in reporting the status of aggregation that could lead to the purified product from being reported as being at a higher or lower level (34).

Purification:

The process of purification processes for biopharmaceuticals can be steered using SEC especially with the reported numbers of biopharmaceutical new chemical entities (NCE) showing an increase. Such molecules are purified initially using affinity techniques employing Protein A or G. Despite the technique offering elimination of impurities from media, the process does not reduce the aggregates formed instead, the process of elution increases the amounts of aggregates formed (Cromwell et al, 2006). This entails the use of additional techniques such as ion-exchange or gel filtration or SEC to reduce such aggregates with the latter offering advantages such as lesser duration involved and the usage of a vast array of buffers that can be applied to the system.

Polymer industry:

For the polymer industry, SEC yields critical information about the chemical composition and distribution of molar mass along with the construction of the molecule. This data is the starting point for properties of a material under study. It entails the use of generating calibration curve using standards of a range of known molecular weights (polystyrene with a low dispersity is a ubiquitous source) (1).

CASE STUDIES

Analysis of Plant Metabolites

A latest 2017 study examined the use of SEC in the analysis of plant metabolites using *Arabidopsis thaliana*. The experiments designed used a principle of the fractionation of metabolites that form stable complexes with proteins simultaneously. This causes such a complex to elute in the the high molecular weight fraction while unbound metabolites eluting in the low molecular weight fraction.

Following lysing of *Arabidopsis thaliana* cell cultures, the supernatant with metabolites was applied to a size filtration spin column using a size cutoff of 10 kDa. Following washing, metabolites were eluted using heat and re- extracted to be subjected to subsequent LC/MS. The authors reported 81 metabolic features that were annotated and reported as cyclic nucleotides (cGMP, cAMP, cCMP), co-factors (FAD, NAD, FMN), and dipeptides.

Size exclusion chromatography (SEC) was then applied to to the native protein-metabolite extract using a column with a range of molecular complexes from approximately 600 kDa to 10 kDa. The reproducible separation of the various complexes was confirmed by a chromatogram of the absorption. The major amounts of metabolic features were detected as unbound with proteins with specific elution profiles of small molecules. 51 of the annotated metabolites identified as binding to proteins from the size filtration experiment were reported in the SEC data. Metabolites that interact with proteins including co-factors such as FMN or NAD that showed distinct peaks in the separation range that indicated protein binding.

The authors provided evidence that *Arabidopsis thaliana* cultures contain a vast array of small molecules/metabolites which form stable complexes with proteins that are indicative of such small molecules with a signaling/regulatory function (35).

Separation of Biomolecules

Another recent work in 2016 listed advantages of using SEC in the analysis and separation of biomolecules. The authors summarized two modes of separation: institution separation where small molecules are sequestered from huge molecules while high-resolution fractionation is related to separation of molecules with alike sizes and dimensions. It was reported that institution separation can rely on smaller bed heights of 10-20 cm with advantages such as rapidity and high throughput. The fractionation mode required the use of longer bed heights in the range of 60 cm to 120 cm along with lesser amounts of sample volumes.

The authors used Column hardware (Superformance 600-16, Superformance 600-26, Superformance 1000-50, Super Compact 240-5) with UV detection on ÄKTAexplorer 100 chromatographic system. They separated mixtures of 2 proteins with molecular weights in the range of 20 kDa and 40 kDa: ovalbumin and conalbumin with masses 45 and 40kDA respectively and chymotrypsinogen A and Trypsin inhibitor with masses 25 and 21.5kDA respectively.

The team also reported the successful purification of a monoclonal antibody using 52.5 x 2.6 cm i.d. (packed to 18% compression) column. Following affinity using Protein A, the sample was subjected to SEC with a removal of aggregates reported to yield the monomer with a purity of 99.5% (36).

Clinical Application: Evaluation of Heparin

Heparin finds use as an anticoagulant for the treatment and prevention of various thrombotic diseases. The analysis of such glycosaminoglycans (GAG) entails depolymerization by either enzymatic or chemical digestion of the GAG polymer to enable structural or separation studies. Such depolymerization steps involve the use of heparin lyase enzymes from Flavobacterium heparinium or nitrous acid or an organic nitrate to chemically attack the structure.

A 2015 study reported the use of SEC for the first time in the quantification of heparin in pharmaceuticals. The of analytical or preparative scale size-exclusion chromatography to separate depolymerized GAGs apart from determining the molecular mass distribution of heparin. The team developed a sensitive method of using SEC for quantitative determination of heparin without depolymerization.

It was also reported that though Azure A colorimetric method for estimation of heparin finds wide use it has certain roadblocks such as its selectivity. The sensitivity of the technique is also compromised by the presence of sodium dodecyl sulfate, bases or salts due to Interference. The SEC method developed was not influenced by such conditions that is an important point as SDS and salts find use in buffers or excipients in drugs.

SEC had the advantage of selectivity as compared to the colorimetric method to assay heparin without the need for grueling preparation of samples. In addition to precision, accuracy the method was validated according to the ICH guidelines. It was successfully applied to evaluate levels of intact heparin in commercial pharmaceutical heparin products and in the experimental formulation of heparin/chitosan nanocomplexes.

CONCLUSION

SEC was reported as a tool for analysis ever since the advent of recombinant proteins: the technique has been applied especially in the field of biotherapeutics to examine the levels of aggregates. The technique also can be used in the study of polymers, guiding the process of purification apart from its role in studying cell lines and manufacturing. The major highs of the technique include simple setups, accuracy, reproducibility and rapidity. Latest techniques have resulted in an enhancement of the sensitivity by the use of small scale particles.

A constant development of the technique can yield an enhanced sensitivity and further augmentation of the throughput to find newer applications in the separation, identification and quantification of various analytes.

REFERENCES

1. Edouard SPB, Koza SM. Advances in size-exclusion separations of proteins and polymers by UHPLC. Trends Anal Chem. 2014; 63: 85–94.
2. Hong P, Koza S, Edouard SPB. Size-Exclusion Chromatography for the Analysis of Protein Biotherapeutics and their Aggregates. J Liq Chromatogr Relat Technol. 2012 Nov; 35(20): 2923–2950.
3. Synge RLM, Tiselius A. Fractionation of Hydrolysis Products of Amylose by Electrokinetic Ultrafiltration in an Agar-Agar Jelly. Biochemical Journal. 1950;46:xli.
4. Barrer RM. General and Physical Chemistry. III. Zeolites as adsorbents and molecular sieves. Annu. Rep. Prog. Chem. 1944;41: 31–46.
5. Wheaton RM, Bauman WC. Non-Ionic Separations with Ion

Exchange Resins. Annals of the New York Academy of Sciences. 1953; 57(3):159–176.

6. Clark RT. Separation of Glycerol from Polyhydric Alcohol Mixture by Nonionic Exclusion. Anal Chem. 1958; 30(10):1676–1678.
7. Lindqvist B, Storgårds T. Molecular-Sieving Properties of Starch. Nature. 1955; 175:511–512.
8. Lathe GH, Ruthven CRJ. The Separation of Substances on the Basis of Their Molecular Weights, Using Columns of Starch and Water. Biochem J. 1955;60(4):xxxvi.
9. Porath J, Flodin P. Gel Filtration: A Method for Desalting and Group Separation. Nature. 1959; 183:1657–1659.
10. Porath J. Gel Filtration of Proteins, Peptides and Amino Acids. Biochimica et Biophysica Acta. 1960;39(2):193–207
11. Porsch B. Epoxy- and diol-modified Silica: Optimization of Surface Bonding Reaction. J Chromatography A. 1993;653(1):1–7.
12. Moore JC. Gel permeation chromatography. I. A new method for molecular weight distribution of high polymers. J Polym Sci Part A: Gen. Pap.1964; 2; 835–843.
13. Czok M, Guiochon G. Aligned Fiber Columns for Size-Exclusion Chromatography. Journal of Chromatography A. 1990;506:303–317.
14. Li Y, Tolley HD, Lee ML. Preparation of Polymer Monoliths that Exhibit Size Exclusion Properties for Proteins and Peptides. Analytical Chemistry. 2009; 81(11):4406–4413.
15. Size Exclusion Chromatography. Principles and Methods. GE Healthcare Life Sciences.
16. Grznárorá G, Polakovič M, Ačai P, Görner T. Extra-Column Dispersion of Macromolecular Solutes in Aqueous Phase Size-Exclusion Chromatography. J Chromatography A. 2004; 1040(1):33–43.
17. Christian J. Evaluation of the "Effective Volume Shift" Method for Axial Dispersion Corrections in Multi-Detector Size Exclusion Chromatography. Polymer. 1999;40(13):3735–3742.
18. Fountain KJ, Neue UD, Grumbach ES, Diehl DM. Effects of Extra-Column Band Spreading, Liquid Chromatography System Operating

Pressure, and Column Temperature on the Performance of Sub-2-μm Porous Particles. J Chromatography A. 2009; 1216(32):5979–5988.

19. Bond MD, Panek ME, Zhang Z, Wang D, Mehndiratta P, Zhao H, Gunton K, Ni A, Nedved M. L., Burman S., Volkin Evaluation of a Dual-Wavelength Size Exclusion HPLC Method with Improved Sensitivity to Detect Protein Aggregates and Its Use to Better Characterize Degradation Pathways of an IgG1 Monoclonal Antibody. J Pharm Sci. 2010;99(6):2582–97.

20. Diress A, Lorbetskie B, Larocque L, Li X, Alteen M, Isbrucker R, Girard M. Study of Aggregation, Denaturation and Reduction of Interferon Alpha-2 Products by Size-Exclusion High-Performance Liquid Chromatography with Fluorescence Detection and Biological Assays. J Chromatography A. 2010; 1217(19):3297–3306.

21. Ahrer K, Buchacher A, Iberer G, Josic D, Jungbauer A. Analysis of Aggregates of Human Immunoglobulin G Using Size-Exclusion Chromatography, Static and Dynamic Light Scattering. J Chromatography A 1009. 2003(1–2):89–96.

22. García MC. The Effect of the Mobile Phase Additives on Sensitivity in the Analysis of Peptides and Proteins by High-Performance Liquid Chromatography-Electrospray Mass Spectrometry. J Chromatography B. 2005;825(2):111–123.

23. Lazar AC, Wang L, Blättler WA, Amphlett G, Lambert JM, Zhang W. Analysis of the Composition of Immunoconjugates Using Size-Exclusion Chromatography Coupled to Mass Spectrometry. Rapid Commun Mass Spectrometry. 2005; 19(13):1806–1814.

24. Kamberi M, Chung P, DeVas R, Li L, Li Z, Ma X, Fields S, Riley CM. Analysis of Non-Covalent Aggregation of Synthetic hPTH (1–34) by Size-Exclusion Chromatography and the Importance of Suppression of Non-Specific Interactions for a Precise Quantitation. J Chromatography B. 2004; 810(1):151–155.

25. Arakawa T, Philo JS, Ejima D, Tsumoto DK, Arisaka F. Aggregation Analysis of Therapeutic Proteins, Part 1. BioProcess Int. 2006; 4(10):32–42.

26. Ejima D, Yumioka R, Arakawa T, Tsumoto K. Arginine as an Effective Additive in Gel Permeation Chromatography. J Chromatogr A. 2005; 1094(1–2):49–55.

27. Ricker RD, Sandoval LA. Fast, Reproducible Size-Exclusion Chromatography of Biological Macromolecules. J Chromatography A. 1996;743(1):43–50.
28. Diederich P, Hansen SK, Oelmeier SA, Stolzenberger B, Hubbuch J. A Sub-Two Minutes Method for Monoclonal Antibody-Aggregate Quantification Using Parallel Interlaced Size Exclusion High Performance Liquid Chromatography. J Chromatography A. 2011; 1218(50):9010–9018.
29. Cordoba-Rodriguez R. Aggregates in Mabs and Recombinant Therapeutic Proteins: A Regulatory Perspective. Biopharm International. 2008; 21(11):44–53.
30. Staub A, Guillarme D, Schappler J, Veuthey J-L, Rudaz S. Intact Protein Analysis in the Biopharmaceutical Field. J Pharm Biomed Anal. 2011; 55(4):810–822.
31. Engelsman J, Garidel P, Smulders R, Koll H, Smith B, Bassarab S, Seidl A, Hainzl O, Jiskoot W. Strategies for the Assessment of Protein Aggregates in Pharmaceutical Biotech Product Development. Pharm Res. 2010;1–14.
32. John S. Philo Is Any Measurement Method Optimal for all Aggregate Sizes and Types? AAPS J. 2006;8(3):E564–71.
33. Rosenberg AS. Effects of protein aggregates: an immunologic perspective. AAPS J. 2006; 8: 501–507.
34. Robbins DC, Hirshman M, Wardzala LJ, Horton ES. High-Molecular-Weight Aggregates of Therapeutic Insulin. In vitro Measurements of Receptor Binding and Bioactivity. Diabetes. 1988; 37(1):56–59.
35. Veyel D, Kierszniowska S, Kosmacz M, et al. System-wide detection of protein-small molecule complexes suggests extensive metabolite regulation in plants. Sci Rep. 2017;7:42387. doi:10.1038/srep42387.
36. Stein A, Heinen-Kreuzig A, Kiesewetter A, Schwarm S. Size-Exclusion Chromatography for Preparative Purification of Biomolecules. BioPharm Int. 2016; 29,4: 30–37, 52.

CHAPTER 13

CHROMATOGRAPHY: THE FUTURE

HOW DOES IT LOOK?

The technique of chromatography is based on straightforward principles for the separation of several samples as discussed in the preceding chapters. The operation of the technique requires a detailed comprehension of the principle, instrumentation and the working of such instruments. However, it has emerged as a strong contender as an analytical as well as preparative technique in the fields of biochemistry, pharmacy as well as biotechnology. The field of multi-dimensional chromatography is expected to see a surge in applications.

The technique originated from the works of Tswett who resolved plant leaf extracts using solvents on calcium carbonate and published the term "chromatography". This was followed by several researchers such as Kuhn and Lederer who were working on animal systems. With initial focus on biochemical applications, the technique has been extended to several fields of science.

The technique has also been applied for compounds that have low molecular weights. Initially, the technique of applying high pressure for such samples that was adding to the cost of the operation due to highly sophisticated instrumentation. This led to the development of use of fine particles that present challenges in the form of packing. Additionally, there is a heat effect associated with low molecular weight substances. Research unearthed the use of Monolithic columns with polymers based on silica. This is reflective of the ongoing research for the expansion of applications in the field of chromatography.

OVERCOMING CHALLENGES

The same holds true for samples that are of complex natures: such complexes are composed of a host of molecules with several functional groups that impart a host of properties. The use of column chromatography in such an approach is limited: research has yielded a technique of multi-dimensional chromatography. This technique can be applied for samples irrespective of their molecular weights. While flow rates for samples are different, the use of particles of fine dimensions in columns of larger length report a high efficiency.

POINT ANALYSIS

The removal of residues from the instrumentation is a vital step in the pharmaceutical industry. The determination of trace amounts of drugs has been achieved with techniques such as UV spectrophotometry and reverse phase HPLC with UV detection.

The technique of chromatography has also entered the realm of detection of clinical samples such as: parathyroid hormone in samples that usually show high levels of cross reactivity. However, the technique of LC-MS/MS aided in the detection of the hormone that is usually present at very low concentrations with very little cross reactivity reported (1).

THIN LAYER AND PAPER CHROMATOGRAPHY

The technique of Thin Layer chromatography and paper chromatography were initially applied for the qualitative or semiqualitative detection of several compounds. However, the development of detectors that are capable of higher orders of optical scanning can aid in the quantitative detection of samples. Based on the uniform application of the layer of stationary phase or application of samples in a uniform manner as well as use of reagents in a uniform fashion accurate results can be obtained. The application of such an approach was tested successfully for the quantification of pesticide residues showing that the future still holds promise for these techniques (2).

LIQUID CHROMATOGRAPHY

The techniques of purification of proteins, sequencing of amino acids as well as analysis of amino acids along with the sequencing of DNA have seen a huge contribution of Liquid chromatography and electrophoresis. These systems shall continue to evolve the fields of biotechnology, biochemistry, life sciences in general. The future looks at the development of systems that operate on a chip to find application in several fields of science (3).

HPLC AND UPLC: TRENDS AND PREDICTION

Now casting a look at the technique of high performance liquid chromatography (HPLC), the technique has been used since more than half a decade in several approaches: preparative as well as analytical. The technique has constantly developing right from the use of initial particles of dimensions 50-μm to the use of particles of really small dimensions: less than 2 μm. There have also been shifts in the pressure: current values of system pressure stand at 20,000 psi from an early operational pump of 1000 psi. Even the detectors used have huge leaps such as operational computation from an initial simple UV detector unit.

A combination of the approach of the scientists in the field and the update of the instrumentation abreast of technology along with a requirement of regulations all make the process of adopting a new technology in chromatography take more than ten years.

It is projected that HPLC and ultrahigh-pressure liquid chromatography (UHPLC) columns market would continue to receive investments in accordance with the present trend. There are reports of 200 business establishments associated with the business with columns listed as consumables. 7 columns annually are used by an instrument that implies increase in the number of columns cause an increase in the market. The overall market for columns (analytical, preparative, capillary-nano, packings, and column accessories) is currently predicted above billion dollars with an overall growth of 3.5% in HPLC but a greater growth in the UHPLC arm.

Recently, the 2.6-2.7 μm superficially porous particle (SPP) columns have hooked up themselves because the favoured column kind for new method improvement in comparison to the sub-2-μm absolutely porous particle (TPP) columns. The SPP present certain materials with a decreased application of pressure with an increase in the efficiency. According to a 2014 Pittcon report, SPP column introductions outnumbered TPP introductions by a tenfold factor.

It is expected that these new modifications would replace the existing materials.

Over the next few years, expectations are that column manufacturers will continue to take advantage of this technology by filling out their SPP offerings with desk bound phases that chromatographers practice

Even with all the current improvements in column particle generation, work is persevering with on in addition enhancements in SPPs. Current strategies for making SPPs are based totally on a multilayer technique or a coacervation process with majority of the SPPs produced in this fashion. There are options being investigated that result in progressed TPP and SPP materials through allowing the formation of more-uniform mesopores compared to present packings. There are reports of silica in the presence of cationic or nonionic surfactants to adopt pores or a two-dimensional method that uses direct synthesis. A More uniform structure has been reported that uses a technique of pseudomorphic transformation (PMT) **(4)**.

GC-MS AND LC-MS

The field of Liquid chromatography (LC) has seen high levels of separation along with the coupling of the technique to LC-MS. This is aiding the evaluation and studies of a cell with more detail: right from the analysis of biomarkers to the complete metabolome. The metabolome of an organism is extremely complex that requires the use of highly sensitive techniques. It is predicted that the field of biomarker discovery shall see a testing of a proposed hypothesis, to perturb a system of characterized metabolites using conditions of disease or alternate culture conditions.

This shall yield a panel of several metabolites that can be used as biomarkers. This system is predicted to use LC-MS in such discoveries.

A combination of LC-MS data with that of other analytical techniques can yield data on the metabolism of organisms.

The coming years are expected to see a surge in collaborations between labs to amalgamate and extract metabolite data derived from LC-MS techniques. LC-MS would see an integration with NMR to evaluate the metabolome and metabolites with the use of offline analysis or a combination of the instruments in tandem. It is expected that GC-MS would be a tool for quantification of samples that have low potentials of ionization and hence cannot be run on an LC-MS system (5).

A MARKET REPORT

According to a 2016 future market insights, the instrumentation of chromatography market is one with huge potential. This is contributed to by several trends such as:

- A surge in the numbers of pharmaceutical and biotechnology Institutes.
- A boost in investments from Governments in the field of R & D
- A rise in the collaborators of several organizations that work with the chromatography market.
- The growth of this market is recording this growth due to the use of these instruments for purification of antibodies and the development of materials of small scales. The market has seen a dip in the numbers of instruments sold due to an economic slowdown as well as the high costs of the instruments. This warrants the need of trained personnel who can operate certain instruments that can further cause a standstill in the pattern.

Across the globe, the largest player in the market is North America with the European region standing second. An increase in funding from the Governments along with a rise in the conferences in chromatography and R&D in the field of drugs is propelling these giants.

It is predicted that though North America would show a dip in the market due to maturity, Asia Pacific is predicted to record a growth. An increase in companies associated with the chromatography market in the countries India, China, Japan, Singapore would add to the trend of growth and expansion.

REFERENCES

1. Grebe SK, Singh RJ. LC-MS/MS in the Clinical Laboratory – Where to From Here? The Clinical Biochemist Reviews. 2011;32(1):5-31.
2. Melvin E. Getz. Past, Present, and Future Application of Paper and Thin-Layer Chromatography for Determining Pesticide Residues. Advances in Chemistry. 1971; 104: Chapter 8: 119–131.
3. Regnier FE, He B, Lin S, Busse J. Chromatography and electrophoresis on chips: critical elements of future integrated, microfluidic analytical systems for life science. Trends Biotechnol. 1999 Mar;17(3):101-6.
4. **Majors RE.** Future Needs of HPLC and UHPLC Column Technology. LCGC Europe. 2015; 28(12): 658–665.
5. Metz TO, Zhang Q, Page JS, et al. The future of liquid chromatography-mass spectrometry (LC-MS) in metabolic profiling and metabolomic studies for biomarker discovery. *Biomarkers in medicine*. 2007;1(1):159-185. doi:10.2217/17520363.1.1

INDEX